AF586922

Student's Companion

Pharmaceutical Marketing Management

Student's Companion

Pharmaceutical Marketing Management

Dr. Bhavna Kumar
(B. Pharm, M.Pharm, PhD)
(PG Diploma in Pharmaceutical Management)
(PG Diploma in DRA, QAQC, Production Management)
Associate Professor,
Faculty of Pharmacy
DIT University Dehradun,Uttarakhand, INDIA

Dr. Abhijeet Ojha
Principal
Amrapali Institute of Pharmacy and Sciences Haldwani

Dr. Anuj Nautiyal
Associate Professor in School of Pharmaceutical Sciences,
Shri Guru Ram Rai University,
Dehradun, Uttarakhand

PharmaMed Press
An imprint of BSP Books Pvt. Ltd
4-4-309/316, Giriraj Lane,
Sultan Bazar, Hyderabad - 500 095.

Pharmaceutical Marketing Management
by Dr. Bhavna Kumar, Dr. Abhijeet Ojha and Dr. Anuj Nautiyal

Published by:

PharmaMed Press

An imprint of BSP Books Pvt. Ltd.
4-4-309/316, Giriraj Lane, Sultan Bazar, Hyderabad - 500 095.
Phone: 040-23445688; Fax: 91+40-23445611
e-mail: info@pharmamedpress.com
www.pharmamedpress.com/pharmamedpress.net

ISBN: 978-93-95039-72-7 (Hardback)

Dedicated to

Father of Pharmacy in India

Prof. MAHADEVA LAL SCHROFF

Preface

Welcome to the world of pharmaceutical marketing management! This comprehensive textbook has been carefully crafted to provide you with a solid foundation in the concepts and fundamentals of this field. Our aim is to not only educate but also inspire you to delve deeper by exploring additional resources.

Keeping in view of all in mind, the authors provided all the efforts to generate initial level of interest by making a better understanding of the fundamentals in marketing. This book will definitely provide an insight on marketing management concepts and can be used successfully in college courses and practical training thatrequire an overview of the critical aspects of marketing management and marketing strategy development.

The book is useful at Diploma, B. Pharm, M. Pharm education levelsas well as forpost-graduate curriculum not only relating to pharmacy profession but also helps for allied sciences who have relevant concepts.In addition to this, the book is useful to the industry, researchers, teaching faculty, doctorates those who are new to the concepts.

Our content aligns with the PCI syllabus while going beyond the prescribed curriculum to include additional topics such as

- Marketing aspects related to consumers.
- Product Design which states the prospective of product and market
- Promotion Pharmaceutical Marketing
- Pricing of Product, Service and concept

We want to emphasize that this book aims to foster independent thinking rather than spoon-feeding information. We have made every effort to provide accurate and relevant information, but we welcome your feedback. If you come across any errors or deficiencies, please do not hesitate to bring them to our attention for consideration in future editions.

We hope this textbook will serve as your guide, empowering you to navigate the exciting world of pharmaceutical marketing management with confidence.

- Authors

Contents

Chapter2

Product Decision

Chapter3

Promotion

Chapter4

Pharmaceutical Marketing Channels

Chapter5

Pricing

CHAPTER 1

Marketing

1.1 Introduction

In our daily routine many products are included as Toothpaste, Soap, Bread, Milk, Car, Mobile Phones, Laptops, and Clothes etc. but from where these products come and where it goes? These products come from the market and their suppliers are known as the ***marketers*** and their receivers are known as ***customers*** or ***consumers***.

These marketers engage in a variety of activities to increase demand for their goods and services in order to generate profit. These activities are referred to as marketing activities because they significantly meet customer needs and requirements in order to facilitate the exchange of goods and services between producers and consumers.

The Latin word "mercatus," which meaning to trade, is where the word "market" originates (To purchase and sell of goods). In a nutshell the meeting of buyer and seller, price determination and transfer of title are the activities, essential for the existence of market.

The definition of market can be viewed by different dimensions:

- The market is the place where buyers and sellers exchange ownership and where products, services, and other objects are bought and sold.
- Market refers to the location where buyers and sellers congregate to enable the exchange of products and services.
- The market can be share market, vegetable market, gold market or geographic market, wholesale market i.e., market dealing in bulk quantity of a particular good or retail market i.e., dealing in small quantities of different goods.
- According to the American Marketing Association, a market is a location where buyers and sellers come together or a hub for certain business operations that control the flow of products and services from the producer to the customer or user.

The first step in marketing begins with the identification of customers' needs and ends with their satisfaction. It entails tasks like designing or merchandising the product, packing, warehousing, transporting, branding, selling, and advertising, as well as upholding positive customer relations to encourage repeat business. Marketing is therefore a social activity in which people trade commodities and services for cash. In order for a company to thrive and flourish, marketing is crucial.

Marketing in the modern era has focused mostly on customer happiness with an eye on any organization's success, which will ultimately play a crucial part in the growth of the nation. The main attention of any firm towards the needs and wants of its customers. Hence, the process of designing, pricing, availability, quality, and promotion etc. of any product are done on the basis of customers' needs.

1.1.1 Marketing: A Definition

- The practise of conducting business operations in a way that facilitates the flow of goods and services from producers to consumers is known as marketing. The traditional definition of marketing was very limited, only included actions linked to moving goods from producers to consumers.
- The social process of marketing enables people to freely make offerings and exchange goods and services with others in order to satisfy their needs and wants.
- Marketing is a collection of processes for developing, communicating, providing, and exchanging goods and services with customers, according to Philip Kotler.

1.1.1.1 Features of Marketing

These are the primary attributes of marketing:

Needs, wants and demands: One of the main objectives of the marketing process is to determine what the target market wants, needs, and demands. To accomplish this goal, all necessary marketing initiatives are made.

Need, finding out what the target market wants, needs, and demands is one of the key goals of the marketing process. The purpose of all necessary marketing efforts is to achieve this goal. It depends on situation when the need of things arises. Example, need of oxygen became important in Covid19 pandemic, at the time lot of oxygen demand was generated for the patient's survival. Needs become wishes to fulfil requests that are made for certain items that could meet the consumer's need.

Creating a market offering: Any business can satisfy customer wants by offering a package of advantages that will entice customers to buy their goods or use their services. Market offering refers to making an offer for goods and services while outlining their characteristics, such as their size, form, quality, and intended uses.

Customer value: Only when a product's or service's values meet their wants in relation to its price will a consumer be willing to buy it. They are getting the most out of the good or service. Value can include elements like pricing, quality, and service. Value rises along with quality and service but falls along with cost. It can be modelled using the formula of:

Value = Benefit / Cost

Exchange: The technique of obtaining a desired product by providing something in return is a key component of marketing. This makes it easier for both buyers and sellers to achieve their goals.

The following conditions must be satisfied before the exchange take place:

(a) The buyer and the seller are the minimum number of parties.

(b) Each party is required to provide the other party with something of value.

(c) The ability of each party to communicate and deliver to the other.

(d) The freedom to accept or reject the other party's trade offer must exist for both parties.

(e) It should be beneficial for both parties to engage in transmission with one another.

1.1.2 General Concepts and Scope of Marketing

Concepts based on which various firms conduct activities of marketing includes:

- **Production concept**: In this concept the company has high focus on increasing the production efficiency and thereby lowering the price of products. Products that are widely available and affordable will be preferred by consumers. Therefore, focus on attaining great production efficiency, cheap costs, and widespread distribution.
- **Selling concept**: This concept focuses on increasing the sale of product by means of advertising and promotional schemes for any company/ organization.
- **Product concept**: Company concentrates on improving the quality of products so as to improve its market sale. Customers will favour goods with the highest levels of performance, quality, or novel features.
- **Marketing concept**: An advanced concept which a company aims to satisfy the needs of consumers. It requires excessive market research and analysis of market behaviour to determine the ideology of consumers. The company manufactures the product accordingly. Target market, customer needs, integrated marketing, and profitability are the four pillars on which the marketing philosophy is built.

Pharmaceutical marketing encompasses pharmaceutical business operations that manage the flow of pharmaceutical goods and services from producers to consumers. Since it encompasses all activities from idea inception through profit realization, marketing has a very broad range of applications. Below is a discussion of a few of them:

(a) **Study of consumer wants and needs**: Products are created to satisfy consumer requests, according to a study of consumer needs and wants. In order to identify the needs and preferences of consumers, research is conducted. These desires and needs act as the consumer's motivating factors while making purchases.

(b) **Study of consumer behaviour**: Marketers do customer behaviour research. By analysing consumer behaviour, marketers may more effectively target and segment their markets.

(c) **Product planning and development**: Product creation, market segmentation, product research, and determining the characteristics, quantity, and quality of the products are all included in this process.

(d) **Branding**: Many reputable businesses use product branding to increase client popularity of their items and for a variety of additional reasons. A choice about branding policy, methods, and implementation programmes must be made by the marketing manager.

(e) **Packaging:** The purpose of packaging is to give the product a container or wrapping for safety, appeal, simplicity of use, and transportation.

(f) **Channels of Distribution:** The marketing manager and sales manager decide which channel of distribution, such as wholesale, distribution, and retailing, is the most suitable.

(g) **Pricing policies:** Marketers must decide on pricing policies for their products. Each product has a distinct price. Among other things, the amount of competition, the product's life cycle, and the marketing goals and objectives are all relevant.

(h) **Sales management**: Identifying customers, determining their needs, persuading them to purchase products, providing them with customer service, and other selling-related activities are all part of marketing.

(i) **Promotion:** Advertising, sales promotion, and personal selling are all examples of promotion. Combining the appropriate promotional activities is crucial for achieving marketing objectives.

(j) **Finance:** Finance is a worry for marketing as well, as each marketing action, such as packaging, advertising, and sales, has a set budget that all operations must be done within.

(k) **Post sales amenities:** Marketing includes providing clients with post-purchase services, fostering positive customer connections, responding to their inquiries, and resolving their issues.

1.1.2.1 Functions of Marketing

Buying: One of the core responsibilities of marketing is the purchase of equipment and raw materials by manufacturers in order to produce their goods. To sell items to retailers and customers, respectively, wholesalers and retailers purchase products from a variety of vendors. It should be tried to obtain the highest quality products from vendors at the most affordable prices in order to maximize profits.

Procedure of buying: Following negotiations with suppliers regarding quantity, quality, price per unit, mode of delivery margins, etc., the purchasing process begins with the preparation of supply orders for the necessary commodities. The following techniques can be used for buying negotiations:

- **(i) By inspection**: This method involves the buyer or his representative physically visiting the supplier's location to inspect the terms, conditions, and product quality. If satisfied preparation of supplier order can be taken.
- **(ii) By sample**: For the buyer's approval, the supplier provides a sample of the product. Buyer evaluates the sample and then creates the supply order in response.
- **(iii) By description:** It speaks of purchases made based on product descriptions seen in a supplier's catalogue or pricing list.
- **(iv) By grade**: By indicating their grade, things are purchased using this way in the necessary quantity (Agmark, ISI, AR, LR, GR etc.)

Selling: It is the second-most significant and crucial aspect of marketing. The processes of buying and selling are connected. It is the procedure used to provide consumers with goods and services. *"Selling is the process of establishing needs, locating buyers, negotiating terms of sale, and servicing purchases,"* claims **Philip Kotler**.

A representation of selling department organization skills has been given in **Figure 1.1.**, while there are various steps involved for selling process as mentioned below.

- **(i) Prospecting:** It refers to targeting the potential buyer and identifying their needs. Such prospective clients can be identified through observations, inquiries, and dealer consultation. Nature and attitude of potential customers must be carefully examined.
- **(ii) Approaching:** To get the customer's attention, the salesperson should approach him. When addressing customers, a salesperson should have a respectful and friendly demeanor. When a salesperson arrives at the counter, he should really welcome the clients. In the event that he is preoccupied with another client, he should reassure the client that he will see him shortly.

(iii) **Presentation:** The presentation's goal is to persuade the buyer that the type of products the seller is providing is what he or she needs. The key characteristics, applications, and unique advantages of the product should be discreetly discussed by the salesperson.

(iv) **Dealing with objections**: Customers may have certain objections, which the salesperson must embrace. These objections are useful for distinguishing the types of clients and their doubts. If the customer has many inquiries and takes a while to decide, the salesperson shouldn't lose patience. But it's important to avoid product-related arguments because they hurt sales.

(v) **Closing the sale:** The salesperson should be approachable when a consumer agrees to buy a product. In order for the customer to believe that his decision to purchase the product was the right one. Even yet, the buyer must be treated decently if he chooses not to purchase the product. The merchant should never pressure a buyer into purchasing a specific product, but he also cannot impose his own opinions on the customer.

Figure 1.1 Diagram of selling department organization skills

Storage: It involves putting in place the appropriate plans to keep the items in good shape from the point of manufacturing until the products are consumed. There are several different types of warehouses that offer storage facilities:

(i) **Private warehouses**: Gigantic group of companies and wholesaler have their private warehouses to store their bulk of stock.

(ii) **Public warehouses:** These are run by companies that offer the general public or small businesses storage facilities to store their things in exchange for a fee. It may occur monthly or yearly.

(iii) **Bonded warehouses**: These are owned by government and situated at airport or seaports. They are licensed to store the imported goods (from other countries) until payment of customs duties. If the goods is purchased in bulk, the corporation may remove it gradually by paying customs duties on the withdrawn portion..

Benefits of storage

- Helps to maintain stability of prices throughout the year otherwise price would be less during season and more during lack.
- Assists in ensuring year-round availability of commodities.
- Modern production happens in advance of demand rather than in response to a specific consumer purchase, therefore businesses need to store huge volumes of things properly.
- Contributes to maintaining a sufficient stock reserve and removing the danger of a delay in getting additional stock of goods.
- Certain products like injectable and chemicals as narcotics requires proper storage facilities to prevent their degradation

Financing: Refers to the procurement of funds for meeting the various expenses of marketing in an effective manner. Two types of funds are required for a company:

(a) **Working capital**: It is the capital necessary for the acquisition of commodities for resale, the payment of employees and salaries, the extension of credit to consumers, and the payment of transportation and storage costs.

(b) **Fixed capital**: It is the sum of money needed to buy assets like real estate, buildings, machinery, furniture, and other office supplies.

The significance of financing maintains the minimal level of inventory in anticipation of demand (importance of financing). To supervise the manufacture and storage of goods for consumer credit facilities, as well as to pay for the costs of buying, selling, and shipping.

Feedback information: It refers to the information that the top management collects regarding- demand and supply of products, latest market trends, preferable package size, and responses of consumers. It is done by extensive marketing research, there are two types resources as internal resources and external resources getting the feedback information as mentioned in **Figure 1.2.**

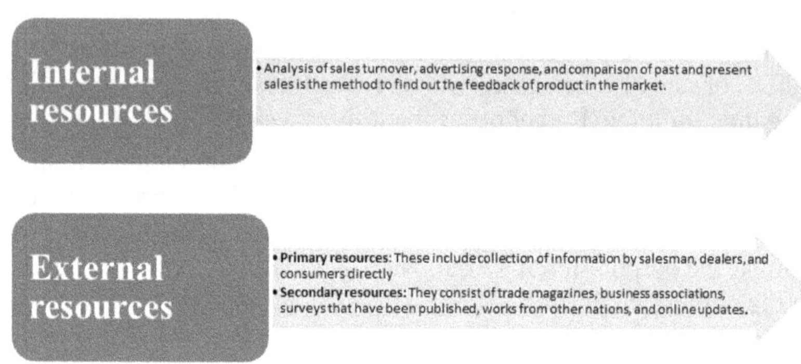

Figure 1.2 Sources for getting feedback information

Significance of feedback information

- It provides details on how customers reacted to the product.
- It aids the manufacturer in learning about competing goods on the market that are similar to theirs.
- It provides insight into potential developments for the product.
- After doing a thorough market analysis, it aids in the introduction of new items.
- Assists in adjusting the product's price structure.
- Provides information on recent government policies affecting a specific business.
- Helps in finalizing the plans to improve sales.

1.1.2.2 Career Opportunities in Marketing

There are various career opportunities in marketing sectors with numerous marketing positions and they are as:

Marketing manager: A marketing manager is in charge of overseeing a company's promotions for its goods and services. Their main responsibility is to increase brand awareness through creative marketing techniques.

Marketing research analyst: Aids businesses in analyzing the market environment for future sales. They are entrusted with providing clients with insights about the serviceability of items.

Advertising or promotions manager: The manager is tasked with planning and overseeing the advertising campaigns and promotions. They coordinate marketing strategies for both new and existing goods and services as well as their promotion.

Product or brand manager: Responsible for developing methods to alter clients' perceptions of a specific product through consumer and trend research. To guarantee consistent branding across all platforms, this may entail supervising marketing and events.

Sales manager: Sales managers have a crucial role in ensuring the organization remains innovative and competitive. They are in charge of establishing quotas, allocating sales training, assigning sales territory, coaching the sales team members, and creating sales plans.

Public relations specialist: Maintains a company's reputation and identity in ways that support its operations, maintains media connections, and works with marketing teams on promotional initiatives.

Marketing coordinator: They are the people in charge of planning all of an organization's marketing objectives and operations. They carry out market research, run advertising campaigns, segment target audiences, and assess current trends.

1.1.3 Distinction between Marketing and Selling (Table 1.1)

Generally, people consider marketing and selling as similar terms but actually both are different concepts. The scope of marketing is much wider than that of selling which is just a part of marketing function.

Marketing is concerned with planning, pricing, promoting, selling, providing after sale services and satisfying the customers. Selling simply means transferring goods and services to customers via promotion, advertising, and sales techniques, which entails turning things into money.

Table 1.1 Distinction features between marketing and selling

S. No	Distinction Features	Marketing	Selling
1.	**Scope**	Marketing's narrow focus is on the transfer of product ownership from vendors to purchasers through advertising and sales techniques.	Selling has a broader focus and is involved with determining the needs of the consumer, creating the product, setting the price, marketing it, and closing the sale.
2.	Focus	Focuses on maximum satisfaction of customers' needs and wants.	Focuses by giving the buyer legal possession of the things and the seller their title.

Table 1.1 *Contd...*

S. No	Distinction Features	Marketing	Selling
3	Aim	Aims for increasing profit by satisfying the consumer.	Selling aims at maximizing the profit by increasing sales.
4.	Sequence	Marketing efforts begin well before the products are produced and continue even after they are sold.	After products are produced, selling operations begin, and they are completed when they are sold.
5.	Emphasis	Under marketing, the emphasis is on developing the product and other strategies according to the need of the customers.	Under selling the emphasis is on sale of products already produced and bends the customer according to the product.
6.	Strategies	Strategies involved related to product promotion, pricing and physical distribution.	Promotion and persuasion are the main strategies under selling.
7	Approach	Integrated approach.	Fragmented approach.

1.1.4 Marketing Environment

The internal, micro, and macro environments that make up the market are referred to as the "marketing environment." It manages the numerous organizational characteristics and conditions that have an impact on the marketing management skills in order to build and sustain favorable relationships with its recognizable consumers.

The marketing environment is a network of external and internal variables that affect a company's capacity to build relationships with and provide for its clients. Among the crucial factors that influence the business environment are industry competitiveness, legal restraints, the effect of technology on product design, and social issues. Every firm needs to consider its operating settings carefully. All businesses must recognize, examine, and keep track of external forces in order to determine how they might affect the firm's products and services. Even though external forces frequently operate outside of the control of the marketing manager, decision makers must take these "uncontrollable" consequences and the components of the marketing mix into account when developing the company's marketing plan and tactics.

"The ability of a corporation to establish and sustain a successful relationship with its target customers is impacted by external variables or forces", **according to Philip Kotler.**

1.1.4.1 Classification of Marketing Environment

Internal Environment: Consists of components found inside or within the organisation. The marketing manager is responsible for creating strategies for the internal environment while considering a variety of groups, including top management, the departments of research and development, finance, manufacturing, sales, and advertising, among others.

Physical resources, financial resources, human resources, information resources, technology resources, etc. are all marketers for the growth of any organization's internal environment. After examining the variables influencing the development of the internal environment, the business must examine the importance of and satisfaction with the client.

Micro Environment: Directly affects the company as it relates to the setting that is most connected to the business. Additionally, business does not have complete control over this environment. An organization's micro environment consists of several forces in its current environment that affect its capacity to function productively in its preferred markets and include:

(a) **Competitors:** There are some fundamental components of the competitive environment that every organisation must be aware of. A firm must deal with several forms of rivalry in the real world of business. The most frequent source of rivalry for a company's product is found in the unique offerings of other businesses. According to Philip Kotler, *adopting the perspective of a customer is the greatest method for a business to fully understand its competitors*. What does the buyer think about the procedure that leads to a purchase in the end? Therefore, understanding consumer sentiment will help all businesses keep their market share.

(b) **Suppliers:** They provide the resources the business needs to generate its products and services. Suppliers play a significant role in the organization's bigger "value delivery system" for customers. Changes at suppliers significantly affect marketing. The marketing manager should keep an eye on supplier availability, delays, supply shortages, and strikes that could briefly boost sales expenses and ultimately have a negative impact on consumer satisfaction. Increasing supplier costs might necessitate raising pricing, which would have a negative impact on the organization's sales volume.

(c) **Marketing Intermediaries:** They help the business market, sell, and transport the goods to the final customers, who might be either people or businesses. They include middlemen (wholesalers, retailers, and agents), market service providers, financial institutions, and distribution firms. Most companies think it's too tough to interact with customers. The distributors and agents help bring the product to the customer in this case.

(d) **Customers:** They give your company the income and cash flow it needs to function and, eventually, turn a profit. You must be aware of and consistently satisfy the wants and needs of your customers. The business must thoroughly assess its target market. Consumer market participants include:

- **Household and individuals:** Buy products and services for one's own usage.

- **Business markets:** Obtain products and services to use in their production process or for further transformation.
- **Resellers market:** Purchase products and services to resell for a profit.
- **Institutional markets:** It comprises facilities that provide goods and services to individuals within their obligation, such as hospitals, schools, jails, and nursing homes.
- **Government markets:** It consists of government entities that purchase goods and services for the underprivileged. Each type of market has distinctive qualities that necessitate rigorous research on the part of the suppliers. The company may at any moment conduct business in one or more consumer markets.

Macro Environment: This environment is made up of a powerful force that affects other macroenvironmental elements in addition to the organisation and the industry. The macroenvironment's components are:

(a) Demographic environment: The market is the first environmental issue that interests marketers because society's members create it. Marketers are especially curious about the size of the society, its density, geographic distribution, age distribution, gender, race, occupation, migration trends, marriage, death rates, and the ethnic and religious makeup of the population.

(b) Economic environment: Economic factors are crucial in determining the purchasing decisions of consumers. Additionally, it has a direct impact on customers' spending power. Even if consumers really like a product or service, they may be unable to purchase it if their purchasing power is low. But if they have cash on hand, they can choose right once to purchase the goods or services they require. The level of consumer income, the income of their family members, spending patterns, credit availability, and other economic factors all have an impact on the decisions that consumers make while making purchases.

(c) Natural environment: Recent years have seen a significant increase in environmental concerns, making the ecological force an important issue to consider. In many parts of the world, water and air pollution have reached deadly levels. The idea that industrial chemicals could rupture the ozone layer and cause the "Green House effect" has sparked a lot of interest. The following four environmental developments present hazards and possibilities for marketers to be aware of: a shortage of raw materials, rising energy costs, increased pollution levels, and dynamic government environmental protection policies. They should contribute to environmental protection by, for example, using renewable energy sources. By doing this, businesses not only help to preserve a green planet but also satisfy consumer demand for environmentally friendly and responsible products.

(d) **Technological environment:** Technology advancements that have an impact on business through new product developments and other advancements in operational methods. Examples of technological advancements can be found in the fields of manufacturing, optics, consumer products, energy, communications, and aviation. To develop procedures and products that satisfy human requirements, technology entails the deliberate use of knowledge, experience, and resources. The many levels of technology include manual, automated, robotic, and computerized. Today, business and technology are essentially interwoven. Among the advantages of technology:

- Shortening the amount of time, it takes to complete a task and
- Increasing the amount of information that can be processed.
- Possessing the capacity to do several jobs concurrently.
- Why Low acquisition costs combined with technology's efficiency and precision reduce the likelihood of error while using technology.

(e) **Political Environment:** It includes laws, government entities, and pressure groups that have an impact on and place restrictions on several organizations and people in a community. There are three main political trends: a sizable quantity of legislation governing industry, fostering the growth of advocacy organizations, and advancing the management of government agencies. Therefore, changes in the political climate have a significant impact on and influence marketing decisions.

(f) **Cultural Environment:** Understanding a person's needs and behaviours depends heavily on their culture. Throughout their lives, a person will be influenced by their family, friends, culture, and society, which will "teach" those people's values, preferences, and norms to their own culture and purchasing habits.

1.1.5 Industry and Competitive Analysis

Industry analysis is also known as *Porter's Five Forces Analysis*. For business strategists, the tool is fantastic. Its foundation is the discovery that different sectors have different profit margins, which may be explained by the structure of an industry.

The main goal of the Five Forces is to determine how desirable a sector is. But the research also offers a framework for developing a strategy and comprehending the environment in which a corporation operates.

The following competitive factors are included in *Porter's Five Forces Analysis* framework, as shown in Figure 1.3.

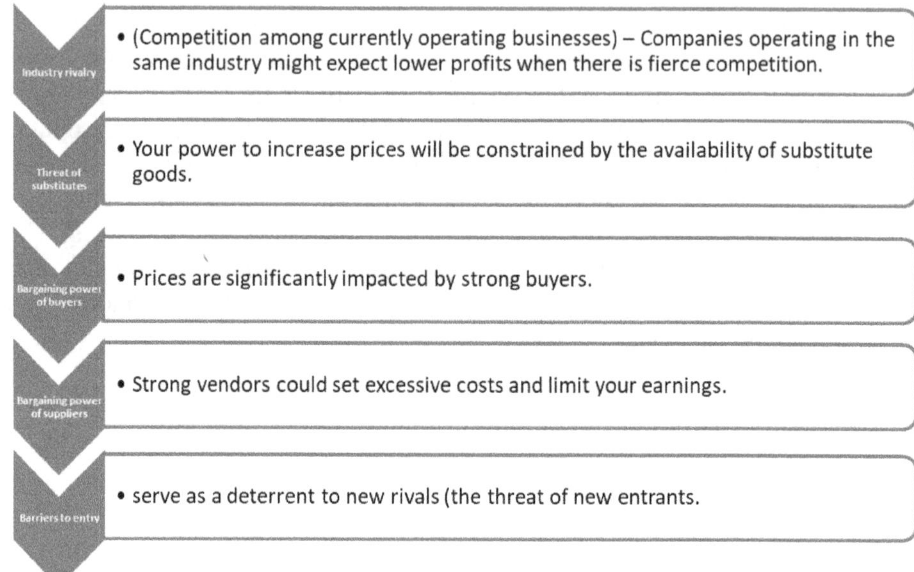

Figure 1.3 Framework of Porter's Five Forces Analysis

1.1.5.1 Industry Analysis and Competition

An industry's internal competition is based on its underlying economic structure. It goes beyond how present rivals are acting. Five fundamental competitive forces determine the level of competition in a given business. The potential for profit in a given industry is determined by the combined strength of these forces. Long-term return on capital invested is used to gauge profit potential. The possibility for profit varies across different industries.

1.1.5.2 Industry Analysis as a Tool to Develop a Competitive Strategy

Through market research, a company can develop a competitive strategy that best withstands the forces of competition or sways them in its favor. Developing a competitive strategy requires an understanding of the origins of the competitive forces. The business will benefit from being aware of these conflicting pressures:

- Highlight the organization's key advantages and weaknesses. (SWOT analysis).
- Animate its position within the industry.
- Decide which areas will be most affected by strategic changes.
- Call attention to areas where market patterns indicate the greatest possibilities or risks exist.

1.1.5.3 Industry Analysis and Structure

The five competitive forces influence each other to determine the level of industrial competition and profitability. The strongest force (or forces), which dominate the market, should be the focus of any industry analysis and subsequent competitive strategy.

The competitive forces that create the basic framework of an industry should be differentiated from short-term elements that have an impact on profitability and competition. Analysis should concentrate on the fundamental traits of the sector, even though these short-term aspects may have some tactical value.

1.1.6 Analyzing Consumer Buying Behavior

Consumer behavior includes all facets of the acquisition, use, and disposal of goods and services. The traits and decision-making processes of an individual have an impact on consumer purchasing behavior. The personality characteristics change with time rather than remaining constant. The study of consumer behavior focuses on how people select, get, use, and then discard goods, concepts, or experiences in order to satisfy their wants and requirements. These psychological traits are dynamic and alter as life does.
Consumer behavior, according to Walters and Paul, "*is the process by which people choose what, when, where, how, and from whom to buy goods and services*".

There are various factors that affect the consumer behavior:

Cultural factors: It tells us about the individual behavior and is composed of values, perceptions, preferences, behavior, nationality, religion, and geographical regions are major factors affecting their behavior.

Social factors: Humans are social creatures because they interact and live with one another. Social factors include reference groups (family, friends, neighbors, co-workers, religious group, professional group). That has direct or indirect influence on the person's behavior. Hence there is a chance of influence of others on their behavior.

Personal factors: Individual factors like age (food habits, clothing habits, differs with age), occupation and economic consideration (affects the attitude towards spending and saving), life style (person pattern of living, depends on its culture, occupation) and personality affects the consumer's behavior.

Psychological factors: The psychology of an individual is based on the level in which he stands in the *Maslow's hierarchy pyramid* (human needs ascending from lowest to highest). Since needs change with every level. The psychology also changes which in turns affects the buying behavior.

Illustration by Maslow's hierarchy pyramid: potential for self-development, realization to fulfil the set goals **Figure 1.4.**

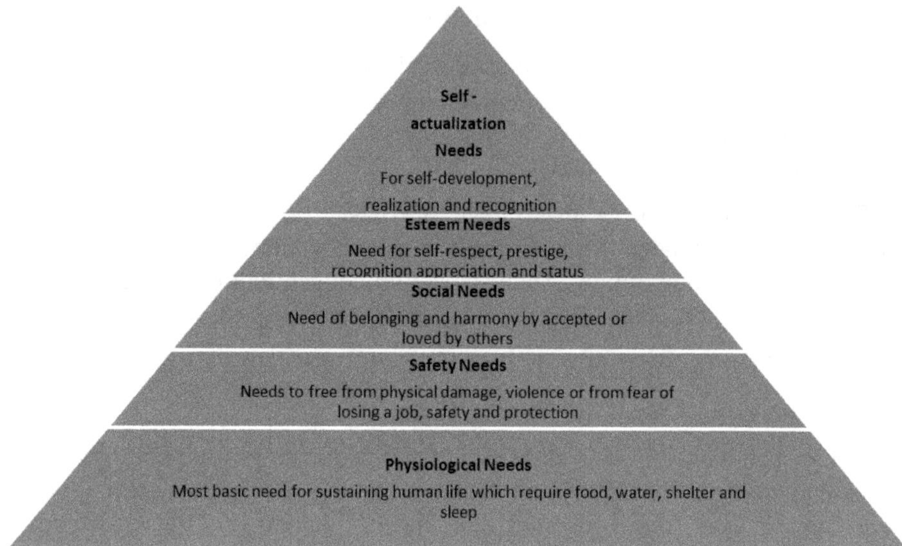

Figure 1.4 Maslow's hierarchy pyramid

The consumer's decision for accepting or buying new products coming in market involves various steps:

Awareness: Consumer becomes aware of products due to its internal and external needs, but lack of deep information about it.

Interest: The consumer expresses interest in learning more about the product via a variety of sources, including family, friends, neighbors', advertisements, salespeople, dealers, and the media.

Evaluation: Consumers evaluate a product's advantages in comparison to those of competing items after acquiring information and then render a final value evaluation.

Trial: A small-scale test of the new product by the consumer yields some amount of satisfaction or discontent.

Adoption: At this point, the customer decides to utilize the product frequently (or else if he is not satisfied, he will not consume it again). The adoption rate of a product by consumer changes according to consumer type:

(i) **Consumer innovators**: Such consumers readily adopt any new product entering the market usually because they are status conscious people (2.5% consumer).

(ii) **Early adopters**: They observe the benefits of product and then accept it (13.5% consumer).

(iii) **Early majority**: This is one of the largest consumer groups. They take some more time than early adopters in critical observation of product (34% consumer).

(iv) **Late majority**: This is another big group of consumers who wants product of good quality but at cheaper rates. So, they take much time to adopt the product (34% consumer).

(v) **Laggards**: They are highly traditional and price conscious people usually they have their own beliefs or attitude so changing their minds is a difficult task (16% consumer).

1.1.7 Industrial Buying Behavior

A corporation engaged in manufacturing, processing, or any other industry exhibits industrial buying behavior. Many of these enterprises must make regular purchases in order to supply their operations.

Although each business and industry will experience its own unique set of influences on buying behavior, there are a number of key elements that can have an impact on industrial buying as a whole, including the following:

(i) **Demand**: Demand is the primary driver of industrial purchasing. The volume of purchases made by an industrial concern directly relates to how much future business the company can anticipate. In general, if a business anticipates more demand, it will stock up on raw resources to make sure it can fulfil customer demand and increase income.

(ii) **Price**: The price of the products that the businesses are buying has an impact on their purchasing behaviour as well. The business may decide to postpone purchases in order to save money when prices are higher or it anticipates a drop in the near future. Making decisions in this situation might be challenging. An attempt may be made to predict the direction of oil prices, for instance, if a corporation uses gasoline in the manufacturing of its product.

(iii) **Economy:** Industrial businesses may look to the economy as a sign of the future availability of materials relative to the consumer demand for them in addition to present demand and prices for a product. If the business is growing, it might make more purchases in anticipation of future sales growth, whereas a declining economy might encourage it to take the opposite course of action.

(iv) **Technological changes**: Technology advancements have a significant impact on industrial enterprises, affecting both their own needs and the ability to provide goods. For instance, if investing in new technology results in a raw material's cost of use decreasing, the business might decide to do so. Similar to this, acquiring new technology can frequently alter the company's purchasing patterns as the new technology would require different raw materials to operate.

(v) **Price increases**: The hope of rising prices is one of the strategic justifications a corporation uses to buy speculative inventory. When a firm has reason to believe that economic forces may increase supplies of commodities or goods, it may purchase more inventory than is immediately necessary or purchase in bulk to take advantage of current market pricing. This is especially likely if the inventory is non-perishable, has no expiration date, and is not prone to degrade over time.

(vi) **Seasonality:** Additionally, businesses purchase speculative inventory to guard against seasonal demand fluctuations. As fall and winter approach, a company that operates in a region containing 4 main climates might buy more snow-related things if it expects a harsh winter. In the event that demand is lower than anticipated, this could lead to surplus inventory on hand, but it also prevents a shortage.

(vii) **Availability:** Lack of manpower and supplies is another potential concern for the merchants that can lead them to accumulate speculative inventory. Customers may stock up on inventory while it is still available if unionised workers in the manufacturing sector are thinking about striking, for example, to guard against a potential loss of supply. Manufacturers of similar products can be worried about how production would be impacted by material loss. For instance, if adverse weather conditions deplete raw materials or there is a shortage of them.

(viii) **Manufacturing**: When customers acquire speculative inventory, manufacturers must also adjust. While materials and products are a priority for purchasers, production efficiency and productivity are a problem for manufacturers. Manufacturers are able to keep up with demand if they anticipate a rise in consumer demand. When customers surprise manufacturers with larger-than-anticipated orders, these businesses may need to hire additional staff, pay overtime, and invest in more resources.

1.1.7.1 Processes Involved in Industrial Buying Decisions

In any process to be run in an industry there is need of buying decisions to be made by the industry which are as follows:

- **Recognition of need:** The business is aware of the demand that can be satisfied by purchasing a good or service.
- **Determining product specification:** The general attributes and necessary quantity of the required item are decided by the buyer.
- **Search for suppliers**: Right now, the buyer is trying to find the best suppliers by using directories, contacts with other companies, trade adverts, etc.
- **Analysis of proposals**: After carefully reviewing the written proposals, the buyers select a small number of the qualified suppliers to participate in a formal presentation.

- **Selection of suppliers**: Buyers will choose suppliers based on the suppliers' reputation, dependability of the products and services, and flexibility of the suppliers.
- **Selection of an order routine**: The purchasers negotiate the final order after choosing the provider, outlining the technical requirements, quantity required, anticipated delivery date, return policy, warranties, etc.
- **Post purchase evaluation:** The chosen supplier's performance is periodically evaluated by the buyer. This evaluation could affect the buyer's choice of whether to keep, modify, or end the supplier contract.

1.2 Pharmaceutical Market

1.2.1 Quantitative and Qualitative Aspects

Statistical analysis, number crunching, huge sample sizes, pre-determined questions, and demographic research are all components of *quantitative market research*. Emails, the phone, and the internet are among examples. Obtaining trustworthy, standardized information and figures is the goal of quantitative research in order to answer important business issues like "Is there a large market for our product?" or "How much does this benefit matter to our target customers?"

Qualitative market research concentrates on customer actions, opinions, wants, and requirements while thinking in detail, building an initial knowledge, and using a small sample size. Qualitative research's goal is to delve further into understanding customer motivation and emotion. Examples include focus groups, one-on-one interviews, and conversations in groups.

There are individual benefits and drawbacks of qualitative and quantitative research as summarized in **Table 1.2.**

Table 1.2 Benefits and drawback qualitative and quantitative research

Qualitative research	Quantitative research
Benefits:	**Benefits:**
✓ Richer knowledge is gained, true motivators are found, and hypotheses are built. ✓ Create hypotheses and concepts.	✓ More effective; Capable of testing hypotheses. ✓ Generalize the findings.
Drawbacks:	**Drawbacks:**
✓ High level of involvement. ✓ Time consuming.	✓ Inflexible. ✓ Lack of motives.

1.2.2 Size and Composition of the Market

The term "market size" refers to the maximum annual volume of sales or customers' business, which is frequently calculated over time. It is helpful to estimate the potential market size before creating a new product line or company line because that can help determine whether the effort and money invested will be worthwhile. A similar concept known as "market share" refers to the full market share that a business has in terms of sales or customers.

Over the next ten years, the domestic market is anticipated to rise three times, according to the Indian Economic Survey 2021. The domestic pharmaceutical market in India is predicted to reach US$ 42 billion in 2021, US$ 65 billion by 2024, and US$ 120–130 billion by 2030. The Indian biotechnology market is anticipated to increase from its 2019 value of US$ 64 billion to US$ 150 billion by 2025. In FY21, India exported drugs and medicines worth US$24.44 billion.

1.2.3 Demographic Descriptions of the Consumer and Socio-psychological Characteristics of the Consumer

India occupies a key position in the global pharmaceutical market. The country also has a big pool of scientists and engineers who could advance the industry to new heights. The Indian pharmaceutical industry meets almost 50% of the global demand for various vaccines and 25% of the demand for all medications in the UK and US. India is the third-largest pharmaceutical producer in the world by volume, but it comes in fifteenth by value. About 3,000 pharmaceutical businesses and 10,500 production facilities make up the domestic pharmaceutical industry.

The major economies of the US, Europe, and Japan face a totally different paradigm when compared to the expanding markets of India, China, South America, and Russia. According to IMS Health, growth in 2009 was 15.9% in emerging markets such as Asia, Africa, and Australia, compared to 5.5%, 7.6%, and 4.8%, respectively, in North America, Japan, and Europe.

Over 80% of the antiretroviral drugs used to treat AIDS (Acquired Immune Deficiency Syndrome) are currently produced by Indian pharmaceutical companies.

This anticipated growth rate is due to India's pharmaceutical industry's favorable macroeconomic climate.

- Following the global financial crisis, India's economy recovered, with real gross domestic product (GDP) growth reaching 9.66% in 2010.
- With rising drug prices and a higher proportion of disposable income going toward healthcare, the Indian middle class is also growing quickly.

- The government has prioritised public healthcare, and as a result, policies and programmes have been introduced to make healthcare more accessible and inexpensive, particularly in rural markets.
- The market is observing trends such as increased merger and acquisition activity, rising investment, deeper inroads into rural and Tier I to Tier VI markets, growth of insurance coverage, and innovation in healthcare delivery.
- The OTC market will be a significant growth engine for the sector. Even though 742 million people, or over 67% of India's population, live in rural areas, just 17% of the market's overall sales come from rural markets. Given that we anticipate these markets to be the industry's future growth engines, this creates a tremendous potential for pharmaceutical businesses.
- Leading Indian and international businesses will seek to enhance their market share by forming strategic partnerships, bolstering their sales teams, and expanding their penetration into additional markets.

The Socio-psychological characteristics of the consumer includes factors which help to improve the industries to improve their turnover and they are as:

Psychological Factors: A key factor in determining consumer behavior is human psychology._Despite being challenging to evaluate, these elements are important enough to influence a buyer's decision. The following crucial psychological elements are:

(i) **Motivation:** A person's purchasing behavior is impacted by their motivation level. A person has many needs, such as social, basic, security, esteem, and self-actualization desires. When such wants are met, people may be inspired to make additional purchases or look for different products and services to better meet those needs.

(ii) **Perception:** Consumer buying behavior is significantly influenced by consumer perception. Sensations are chosen, arranged, and interpreted through the process of perception. It involves obtaining information about a product and then evaluating that information to paint a precise image of that particular product. Customers base their opinions of things on what they see in advertisements, promotions, customer reviews, comments on social media, etc. Customer perception so significantly affects what consumers decide to purchase.

(iii) **Learning:** Learning can only occur through experience. A person can only learn about a good or service after using it. An individual will exhibit a strong propensity to repurchase a given good or service if they are happy with it the first time.

(iv) **Attitudes and beliefs:** Attitudes and perceptions of consumers have a big impact on their purchasing decisions. Every product or service available on the market has a specific image in people's minds. Every brand has a corresponding image, also referred to as its brand image. Consumers make their decisions to buy a product or service on the views they have developed about it. Even though a product might be excellent, if the customer believes it to be useless, they won't buy it.

Social Factors: Social factors have a substantial impact on consumer purchasing decisions. We need people around to talk to and discuss a range of issues in order to develop better solutions and ideas. Since we all belong to a society, it is essential that each person follow its norms and regulations. Numerous societal components include:

(i) **Family:** The family has a big part in influencing a person's purchasing habits. A person forms preference as a youngster by observing their family members purchase goods, and they maintain those choices as they get older. People usually ask their close relatives for advice before purchasing a specific commodity or service. A person's family may urge him to buy a specific thing, ban him from doing so, or make a few other recommendations.

(ii) **Reference Groups:** Everybody is surrounded by people who have an influence upon them. Reference groups are made up of persons to whom people compare themselves. Everyone is familiar with a few community members who, over time, become their heroes. Reference groups are typically created by coworkers, family, friends, neighbors, elders at work, individuals of the same religion or political affiliation, clubs, etc. All of the aforementioned have an effect on customers' purchase decisions for the reasons given below.

(iii) **Roles and status in the society:** Individuals' propensity to purchase depends on the part he performs in society. Each person serves a dual purpose in society, depending on the group to which they belong. A good company's chief executive officer could also be a husband and a father at home. A person from the middle class or a lower income group would buy things they needed to survive.

1.2.4 Market Segmentation and Targeting

The practice of segmenting the market for a product is breaking it down into a number of smaller marketplaces or segments. Segmentation is the process of identifying the customer groups within a market that have comparable needs and buying habits. There are one billion consumers in the globe, each with unique requirements and habits. The purpose of segmentation is to match various buyer groups with various sets of needs and purchasing patterns. Segment refers to such an assortment. The reasons for developing the market segmentation are mentioned in **Figure 1.5.**

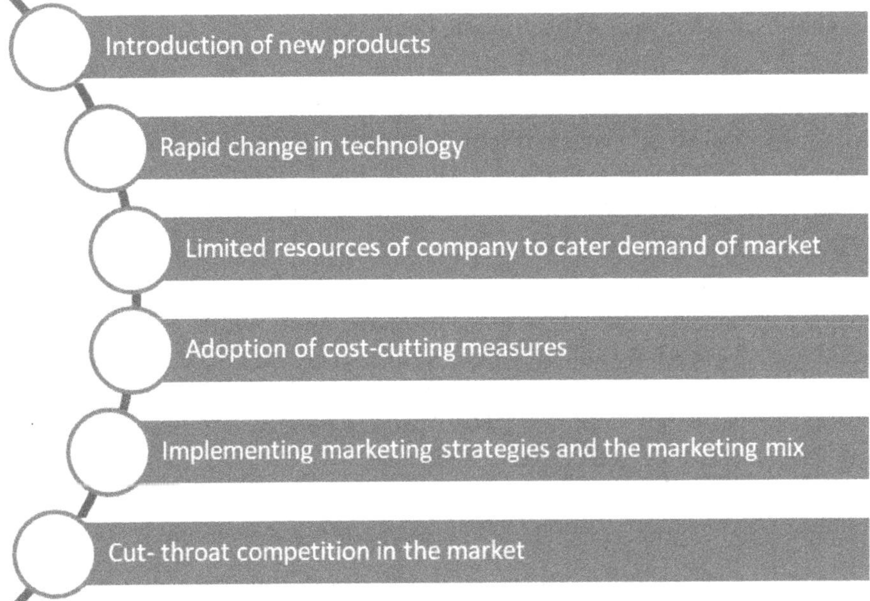

Figure 1.5 Reasons for development of market segmentation

1.2.4.1 Bases of Market Segmentation

***Geographic**:* Many firms divide their markets using a variety of geographic factors, including population density (urban, suburban, and rural), city size (area size, population size, and growth rate), and climate (Regions having similar climate pattern). A business that caters to some or all geographic segments must focus on the diversity of regional needs and preferences. Companies localize their marketing strategies after segmenting the consumer market based on geography (product, advertising, sales efforts, and promotion).

***Demographic**:* It is the most well-liked consumer market segment, and it is determined by a variety of variables including age, gender, earnings, profession, education, size of the family, and religion. Demographic factors are the most important criteria for categorizing clientele into groups. Consumer demands, preferences, and usage patterns are all influenced by demographic factors. Thus, it is essential to take demographic considerations into account while developing a marketing strategy. Dividing the market on the base of stage of age (as infants, children, adults and geriatrics) have different preferences. Examples can be summarized as cases of different requirement of garments for different age groups, medicines generally formulated in sweet syrup form specially for children, as the bases of gender (male and female segment: it has been successfully applied in clothing hair styling, cosmetics, magazines), income (high level income, middle class-upper middle class, lower middle class, low level like below poverty line). Annual income of the population is a basis for segmenting market of various products such as automobiles, cosmetics, travel, clothing etc.

Psychographic segmentation: In this kind of segmentation, characteristics like socioeconomic class, manner of life, and personality qualities identify the groups. Examples of psychographic elements include interests, opinions, personality, self-image, hobbies, values, and attitudes. A subset of demographically defined consumers might exhibit certain psychographic characteristics.

Behavioral towards product: The market is divided into segments based on consumer knowledge, attitude, use, or response to a product. Product, benefit, brand loyalty, occasion (holidays like Mother's Day, the New Year, Pharmacy Day, etc.), and user status are a few examples of behavioral factors (first-time, frequent, or potential). Example: ***Hard core loyal***-Loyal to one brand, ***Split loyal***- Loyal to two or three brands, ***Switchers*** - No loyalty). Using criteria that are closely tied to the product itself, behavioral segmentation is thought to be the most effective segmentation technique.

Table.1.3 Advantages of Market Segmentation

S. No	Advantage to the Firm	Advantage to the Customer
1.	Increase in sales volume and market share.	Customer oriented
2.	Prepare effective marketing plans, to win competition.	Quality products at reasonable prices
3.	Helps to enables to take decisions and understand the needs of consumers	Other advantages (By offering discounts as a buy one, get one free promotion, customers win)
4.	Utilises resources wisely to meet marketing objectives.	
5.	Allocation of marketing budget and create the innovation for developing specialized markets.	

As seen in **Table 1.3**, the market segmentation has its own benefits.

Limitations of Market Segmentation:

(i) Market segmentations shows their limitations due to extraordinary cost on production because of high advertising and promotional costs.

(ii) It also shows hike in working capital due to high administration expenses and storage expenses.

(iii) Challenge to choose variable segments and striving in getting skilled and experienced workers.

1.2.4.2 The Advantages of Market Segmentation

(i) **Recognizing customer demands**: Aids in the marketer's ability to thoroughly comprehend consumer demands, behaviour, and expectations.

(ii) **Allocation of marketing budget**: Aids in distributing the marketing money to a specific area or locale. For instance, a modest budget should be set aside in areas with fewer sales chances.

(iii) **Effective marketing programmes**: The manufacturer can create and implement various marketing strategies for various segments based on segmentation to ensure effectiveness.

(iv) **Increase in sales volume**: By using segmentation, the manufacturer may better understand each segment's demand patterns and create the appropriate items to meet their needs. As a result, the product's entire sales volume rise.

(v) **Benefits to customers**: It benefits the customers because by using segmentation manufacturers can know the demand of each segment and can make products which serve the interest of customers.

(vi) **Encourage innovations**: Marketers gain advantages from segmentation in that they can focus on the pertinent segment more intently and monitor changes in market demand.

1.2.4.3 Significance of Segmentation

There are few criteria for successful segmentation as mentioned in **Figure 1.6.**, and their segmentation have some significance in the market.

- To comprehend consumers' needs, wants, desires, and purchasing patterns.
- Assists the advertiser in understanding consumer behaviour, preferences, and nature.
- To create marketing strategies tailored to each consumer group's unique needs.
- Create new items in response to shifting consumer demands.
- Create expansion of market and market share.
- Marketers can open up new markets for their goods.
- Market segmentation is first performed which refers to dividing the customers based on
- Provides satisfaction to consumers.
- Allocate appropriate budget which is assigned for a particular region.
- To make marketing tactics and regulations customer-focused.

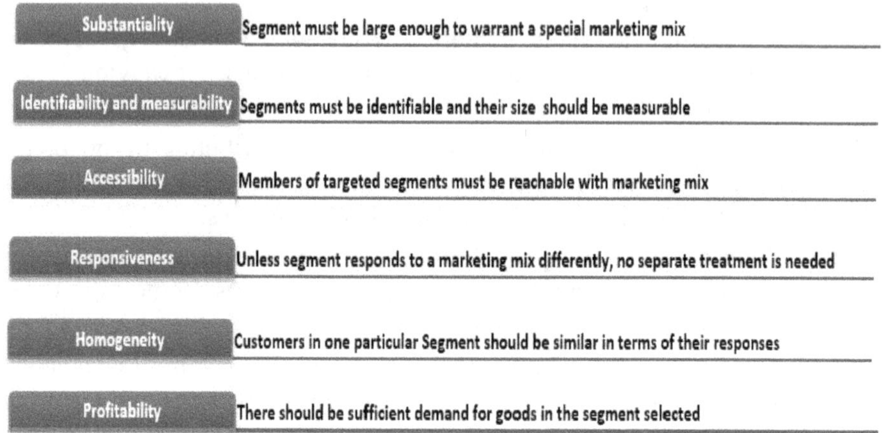

Figure 1.6 Criteria for successful segmentation

1.2.4.4 Target Marketing (Target consumer profile)

Target market: It consists of a group or groups of consumers that the business seeks to please or for which products are produced, prices are established, marketing initiatives are undertaken, and distribution networks are set up OR a target market is a collection of consumers that the company's whole marketing efforts are aimed towards.

In target marketing, a marketer selects one or more of the identified market segments, then produces goods and other components (such as pricing, location, and promotion) that are suitable for each segment. A collection of people or organizations that an organization creates, develops, and maintains the marketing mix for is referred to as a target market (This term first introduced by Borden in 1964). The four components that make up the marketing mix are what the company utilizes to pursue its marketing goals in the target market. To gain the customer's approval, the factors of product, price, location, and promotion all worked together.

Product: is a good, service, concept, location, or individual that is provided to the client in order to meet his wants. The characteristics of items that must be aimed towards clients are: *Quality, Warranty, Packaging, Design, and Services*.

Price: Is the price that a consumer is willing to pay for a product? e.g., for different social class of customer's company designs various products of different prices. Recharge coupons of mobile come in different price range.

Place: It is the process for delivering products from manufacturers to the intended consumers. This business collaborates with accredited dealers for the aim of product distribution.

Promotion: Incorporates a range of customer targeting strategies, such as advertising, sales promotion, public relations, offers and schemes, and personal selling. after choosing the segmentation basis. The business must specify who or what organization the product was intended for.

1.2.4.4.1 Strategies of Market Targeting

There are three basic strategies that a company may adopt for targeting-

Undifferentiated marketing: In this market coverage strategy, the corporation regards the target market as a single entity and does not think the existence of many segments. With a single offer and a single marketing mix, the company targets the middle of the market. Thus, in undifferentiated marketing company adopts the strategy of "**one product all segments**"

Differentiated Marketing: As part of its market coverage strategy, the organization thoroughly examines the market and separates it into relevant divisions. On the basis of market segment company covers these market segments using different marketing mix for each segment. Hence the company follows the approach of "**Several products- Several segments**"

Concentrated Marketing: The corporation adheres to the ***"One Product - One Segment"*** philosophy in this strategy. The business gathers comprehensive information about a specific market niche and concentrates exclusively on it. Company designs the ideal marketing mix for this segment and develops varieties of products for this segment. This strategy is very helpful for new and small companies to make a good reputation and standing in the market by focusing on a special segment.

There are several factors that govern the selection of target marketing strategy:

- Product homogeneity.
- Resources available with the company.
- The product's position in the product life cycle.
- Market homogeneity.
- Marketing strategies of competitors in market.

Advantages of target marketing:

- It helps the business better capitalise on marketing opportunities.
- In order to better serve its consumers and gain devoted ones, businesses might offer products that are appropriate for each target market.
- It can expand the market share.
- Possibility of improving brand recognition for the organisation.

1.2.5 Consumer Profile

The goal of the consumer profile is to develop a thorough understanding of the customer for marketing and research reasons. It is a compilation of information about the consumer as demographic, geographic, psychographic, and behavioral data. It is well explained in **section 1.2.4.1** under bases of market segmentation.

1.2.6 Motivation and Prescribing Habits of the Physician

Patient motivation refers to the guiding, encouraging and motivating the patients to take the particular therapy in appropriate dose and taking necessary precautions during illness. Motivation of patients is beneficial in following manner.

- It guides the patient at times when any change from normal health has occurred. Some persons have the tendency to avoid the normal health problems that can later turns into serious disease, so motivation encourages persons to be cautious if the illness symptoms are frequent and persistence.
- Patient motivation also helps in improving the decision of patients regarding selection of doctors. Careless patients view consultation fees of doctors and then select the doctor. Proper motivation of patients decreases the tendency of such unfair selection strategy of doctors by patients.
- Some patients go to worthy doctors take their prescription but they purchase drugs by their own experience and not as per prescription. Such practices may cause serious consequences. Patient motivation reduces such incidents by proper guidance.
- Generally, a majority of patents have the tendency to buy the drugs as prescriptions but only in half dose. They think that half dose will show effect but they don't have the knowledge about drug potency. Proper counselling discourages such practice.
- Many patients tend to take the prescribed drug in proper dose but they neglect the necessary guidelines to be followed along with taking drugs. This can cause less or no effect of drugs taken. Hence patient must be regularly and carefully motivated towards required change in their behaviour and psychology related to use of medicines.

1.2.6.1 Prescribing Habits of the Physician

According to **American marketing association (AMA)** A prescription is an order for the pharmacist to prepare and deliver medication for a patient together with instructions for its administration and use that is written down by a doctor, dentist, or other licensed practitioner. The prescribing behavior of physicians is affected by a number of variables which have been reported by AMA.

The following factors influence doctors' prescription behaviour:

(i) **The patient's clinical and behavioural traits:** The patient's demands and expectations for medication-assisted therapy, the treating physician's attitudes, expectations, and training, as well as any organisational or contextual restrictions on the doctor.

(ii) **Research and pressures from pharmaceutical organizations:** Pharmaceutical and research firms are intertwined, and they may influence physicians through scientific publications, instruction, and advertising based on the physician's character attributes, professional background, and treatment options. Pharmaceutical organizations affect physicians through education; scientific journals, advertising and profitability.

(iii) **High curing effect; habitual or non-habitual choice**: The most crucial factor in choosing a medication to treat a condition is always its "curing effect." Drugs are chosen by doctors either based on habit or free choice. Most of a doctor's medicine selections are habitual, and they prescribe a new prescription using a non-habitual approach.

(iv) **Peer influence of physicians (Community of physicians):** The social organisation of a physician community was influenced by peer pressure on medical professionals' prescription. When treating challenging illnesses where the effects of medication therapy are less clearly known, doctors consult more professional sources, especially their peers. A doctor consults a colleague for information about medications that they do not already use. It accounts for 2% of a doctor's prescription behaviour.

(v) **Hospital staff meetings**: Positively affect the acceptance of new drugs, and formularies almost certainly have a big impact on prescribing. It accounts for 2% of a doctor's prescription behaviour.

(vi) **High Cost of Drugs:** Prescribers in impoverished nations frequently utilise expensive medications inappropriately. The lack of a well-organized drug policy and the inadequate economic resources in developing nations make problems there more severe. Strengthening and enforcing drug policy is unavoidable in order to promote logical prescribing patterns and prescription quality.

(vii) **Medical representative:** The medical representatives account for 48% of the total factors that affect the prescription habits of physician. If medical representative is well behaved, knowledgeable, disciplined and visiting the prescriber regularly, there are many chances that physicians will prescribe the drug.

(viii) **Medical Journal articles:** The articles related to new drugs published in medical journals has 30% share on affecting the prescribing behaviour.

(ix) **Direct mailing**: Some companies directly mail the recent information about their drugs to the physician. This affects prescribing habits up to 4%.

(x) **National and international conventions**: The national and international conferences or seminars related to drug awareness is responsible for 2% of the variables affecting prescribing behaviour.

(xi) **Other factors**: Effects 10% of prescription decisions as:

- Physicians frequently prescribe newer, more expensive medications to patients who have more money.
- Consistency and therapeutic response are positive attribute for influencing drug adoption by physician.
- Low cost of therapy is also one of the positive attributes for increasing prescriptions.
- Education and the right training have a favourable impact on the standard of prescribing.
- The type of hospital in which physician does the practice also influence prescribing behaviour.
- Free samples provided to the physician have also some positive influence on the physician prescribing tendency.
- Physician in private clinics usually prescribes more cautiously than in smaller government hospitals. Similarly, large hospitals doctors usually provide quality prescription than smaller village side hospitals.

1.2.7 Patient's Choice of Physician and Retail Pharmacist

By actively participating in treatment decisions, being well-informed, and changing doctors if they feel their care is not up to par, patients can play a significant role in obtaining optimal health. There is widespread agreement that, as the availability of health information online rapidly expands, consumers' influence over the health care delivery system will grow over time.

Factors associated with patient choice of physician:

(i) Patient choice is based on doctor certification and can be also recommended by the family and friends.

(ii) Doctors specialization, appointment time period, proximity of doctor clinic.

(iii) Choice is also based on doctor's fee/cost, seating of doctor (in clinic, hospital, dispensary etc).

(iv) Mostly the choice is based on popularity, consultation hours and orientation of patient towards the patient.

Retail pharmacy is the pharmacy where prescriptions are filled or medications are distributed to the general public, as well as where medications are compounded, dispensed, stored, or sold. Additional explanations of retail pharmacy include:

- It can also be defined as a pharmacy where a customer gets medicine directly as per prescription or
- Unlike a hospital pharmacy, it is a pharmacy where patients can purchase medications. Likewise known as a neighbourhood pharmacy.

Retail pharmacists give the public basic health advice and supply them with both prescription and over-the-counter medications. A retail pharmacist's responsibilities include processing prescriptions, distributing medication, and responding to consumer inquiries regarding health problems, symptoms, and drugs. They are also supervising the medical reps and hiring, training, and managing workers. They are involved in placing orders for and selling medications, maintaining inventories, and handling budgets and financial records.

Steps involved in working of retail pharmacist are:

1. **Accept and check prescription details**: This is for the details required of prescriber, and patient. Medicare will certify the prescriptions to be filled and its preferences for the generic drug.
2. **Review and process:**
   (a) **Script validity**: Should be eligible for the pharmaceutical benefits system and meet the legal requirements.
   (b) **Safety and appropriateness**: To provide a safe dosing regimen and to provide the contra-indications, it is necessary (Inappropriate for people with some medical conditions). The suitability of the prescription requirement based on factors like age, body weight, sex, etc.
   (c) **Review patient's dispensing history**: Examining the patient's medical history, paying particular attention to any new or modified therapies, any duplications, interactions, compliance concerns (is the medication being taken as directed?), uncommon use, and abuse/misuse issues (drug-drug, drug-disease state, drug-herb.
   (d) **Patient-specific factors**: The patient age, various allergies and other health condition.
3. **Select/Prepare and check:**
   (a) **Select product:** On the basis of appropriate drug, brand, strength, quantity a product is selected. The need of repacking if the case is for non-standard quantity. Reconstitute or compound from raw ingredients wherever is required.

(b) **Dispensing check:** For compounded goods, a proper drug, brand, strength, form, and quantity, as well as an exact formula or methodology, will attest to the product's therapeutic advantages.

4. **Label and assemble:** The dispensed items are labeled and reviewed for expiry, instructions, and cautionary labels. Also, the barcode scan check is required for labels. It is necessary to verify all documents, records, and counselling tools (such as written materials).
5. **Supply and counsel:** This will benefit the patient if a correct medicine is given to a right patient as well as the document is correctly documented.

1.2.8 Analyzing the Market

Market analysis refers to understanding of market structures in order to determine the pricing level, nature of demand and supply and marketing strategies. A market study provides answers to a number of questions about the consumers, including who they are, what they need, how they choose products, and where they get them. Additionally, how do you communicate your marketing and sales messages to them?

The structure of market is based on four factors:

Number of sellers: If the number of sellers in market is large the influence of one seller is small on the prices. In case, if market has very few sellers the pricing decision of each seller is important. In such case seller is the decision maker for price and product output.

Numbers of buyers: If buyers are less in the market, they can demand less price for product and thus affect the pricing decisions. On the other hand, if buyers are more they have to pay the same price.

Product differentiation: Product difference is the degree at which one product differs from other in the market. When there is much product differentiation, the decision of buyer is based on price. In such case every company tries to differentiate its product and to create a good impression on the minds on consumer by offering quality product at lower prices.

Condition of entry and exit: The easiness of entry and exit of companies in market decides the market structure, pricing decision and strategy of the companies. If there is a free entry of companies in market (i.e. less restriction to company's approval by government), the competition for existing companies is high and their own decision are affected by competition. If entry of new companies is difficult the existing company can freely make its pricing decisions. If any company incurs loss and resources (raw materials) available can be used in producing other drugs. It can freely exit from market and can start with some other product. In such cases pricing decisions and strategies can be easily changed by the company. On the other hand, if resources available have highly specialized uses and very few alternate uses then exit decision is very difficult. In such case strategies of company must be designed critically.

Need of market analysis: Market analysis provides information of market and identify the potential customers, their need and provide insight into existing customers. The market analysis also offers the customer behavior pattern, and identify the business opportunity. The determination of sales potential of the product and services attracts the customers to the business, which will give a strategic advantage to resolve the business problems.

Cournot classified the market structure as:

Highly competitive market: A market with perfect competition includes elements that demonstrate a high volume of customers and sellers in the market, product is highly homogeneous (less product differentiation) which makes the entry of firms easy in the market. All buyers and sellers have perfect knowledge about market conditions. The pricing decision is not done by company freely. Price is almost fixed and preset for a product and new company will decide the output (product volume) to be produced according to their own economic condition.

Monopolistic market: When number of sellers (company) is very few in the market, it is called monopolistic market. Its essential features are the number of sellers (producers) which are less. No close substitutes are available for product in the market, and entry of new company is restricted due to tough govt. regulations.

Monopsonistic market: Monopsony is a market situation where there are very few buyers (consumers) for products. The features of this market are: Where the number of consumers (buyers) are less, buyers have full control on the price of market because they can easily bargain for the prices. Thus, monopsony aims at maximum surplus. The consumer's surplus is the difference between the prices of product and the amount that consumer actually pays.

Bilateral market: It is a market in which a monopolist faces a monopsonist. It means that less number of buyers as well sellers are available in the market. The features of bilateral market are: Showing the market with very a smaller number of sellers and there are a smaller number of consumers in market. The product in market have no close substitutes, the monopolist will try to maximize his profits while monopsonist will try to maximize his consumer's surplus.

1.2.8.1 Sales Analysis: Analysis of actual sales data is referred to as sales analysis. This is distinct from sales forecasting in that it focuses on current sales performance rather than sales at a later time. A corporation benefits from sales analysis in two ways-

- It allows the business to pinpoint the regions in which its sales performance has been strong or mediocre.

- It is helpful for a company to strengthen its sales effort to do systematic and periodic sales analysis.

Sales analysis is done in three ways:

A. **Sales analysis by territory**: Two things must be chosen in order to perform sales analysis by territory:
   - What specific information is required for such purpose?
   - On which basis territories should be divided for analysing district level and state wise analysis.
   - Generally, most of the companies use district level analysis. In this method district wise analysis is done and after pooling the district data the state wise sales data is prepared. This method of analysis by territory is the most popular.

B. **Sales analysis by product**: It enables a company to identify its strong or weak points. If a company finds that a particular product is showing poor sales, it has two options-
   - The company might focus on that product to assure higher sales.
   - The sales analysis by product and that by region can be integrated, which is useful in providing information on which items are displaying good sales in which locations.
   - The company may progressively withdraw the product and eventually drop it.

C. **Sales analysis by customer**: In this method customers are divided into: Potential customers (buy less but in bulk), first time customers and regular customers (buy regular but in small amount). Then a sale in each class of customers is analyzed.

1.2.8.2 Sales Forecast

Sales forecast refers to the estimation of sales that are likely to be at a future date. Thus, it is a future prediction phenomenon of sales. The forecast depends on the demand pattern of a product in the market. Demand pattern for some products changes over time; it is called the *dynamic pattern* of the demand. On the other hand, if demand pattern retains its predetermined shape then it is called *stable pattern* (It refers to the prediction of the conditions of demand or sales that will occur at some future date. The overall business and economic environment, competitive factors, market trends including changes in demand, organizations' own plans for advertising, product promotion, and price policies, changes in product features or design are just a few of the variables that might affect the forecast of demand. The steps of sales forecasting are mentioned in the **Figure 1.7**

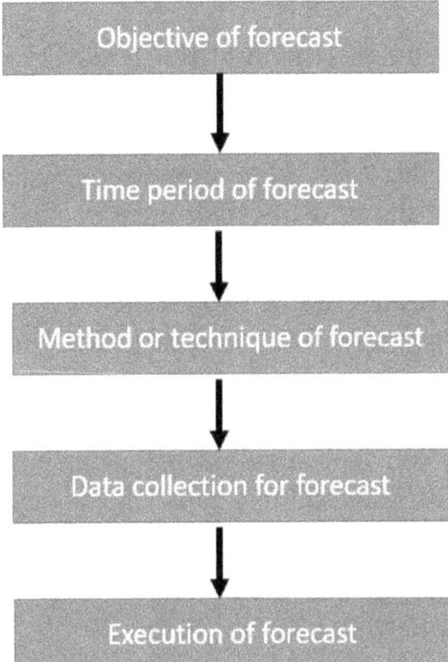

Figure 1.7 Steps of sales forecasting

1.2.8.2.1 Methods of Forecasting

A. ***Qualitative methods:***

Field sales force method: In this approach, the businesses ask their salespeople to forecast the likely sales in a given time frame for their region. District managers review the area-by-area sales before sending them to the company's headquarters. The total sales projection is then obtained by combining several territorial forecasts. The method's relative accuracy is high since it includes the full sales team, which is in charge of making the sales. Thus, the sales team can provide a precise sales forecast for a given period of time.

Jury of executive's method: In this method a company appoints some qualified experts and assigns them the task of sales forecast. The experts predict the sales for a particular time period. They are given an opportunity to meet, discuss their forecasts with each other so as to achieve to a finally approved common sales forecast.

Delphi Method: This method also involves sales forecast by means of qualified and experienced jury experts but this differs from jury method in that the jury experts are not allowed to meet or discuss their forecast. The individual forecast may differ hence average of there is calculated to obtain the final sales forecast. This method reduces the group pressure on each expert otherwise some dominating experts may pressurize others to agree on their forecast.

B. *Quantitative method:*

Time series method: In this method the company collects the past sales data, which are plotted on a curve. The curve is then extra-plotted to get the future sales, otherwise extrapolation can be done using statistical calculations. It is suitable in those conditions when past sales are in a particular trend which shows a linear or curvilinear (somewhat linear) plot, only then extrapolation can be done accurately.

Moving Average method: In this method "N" data points of past sales are used to calculate an average. Then the earliest figure is dropped and each time latest figure is estimated using the average. Or in this method an average of past data is taken usually three or five data are averaged to get the data for next period. To calculate the new average for the following period, the first period's demand data is eliminated, and the most recent period's demand data is added.

Example: A company takes average sales of January, February, March, April and May to calculate average sales.

$$\textbf{Average Sale (S)} = S_{Jan} + S_{Feb} + S_{Mar} + S_{Apr} + S_{May}/5$$

Now, the company can predict sales of June by assuming the average sale to be constant and omitting the $S_{Jan.}$

$$\textbf{Average Sale (S)} = S_{Feb} + S_{Mar} + S_{Apr} + S_{May} + S_{June}/5$$

Once S_{june} is predicted the same omitting method is used for predicting the sales of S_{july}

Example: If demand for past three months has been 120, 135, and 114 units. We can calculate forecast for 4th month.

120+135+114/3 =123 unit

Suppose, the actual demand for 4th month is 129 units, then forecast for 5th month will be:

=135+114+129/3 =126 unit

Demerit: This method assumes that sales average will be common which is practically not possible because sales can change according to some other external factors also like- Buyers interest, buyer's income, market factors, competitor's product, pricing policies. This method does not consider the market factors and change in demand pattern; hence fluctuations may occur.

1.2.9 Market Research

The process of connecting customers or consumers with marketers through the utilization of information is known as marketing research. The information is used to identify and classify marketing opportunities and problems, to develop and refine marketing actions, to monitor marketing performance, and to improve understanding of the marketing process. There are numerous definitions provided by various:

"Gathering, organizing, analyzing, and sharing data that can be used to inform marketing decisions are all part of marketing research".

"The systematic, unbiased search for, and analysis of, data relevant to the recognition and resolution of any problem in the field of marketing constitutes market research".

"Marketing research is the methodical, objective, and thorough search for and analysis of the facts pertinent to any problem in the field of marketing," writes Richard D. Crisp.

Marketing research is "the systematic planning, collecting, analysis, and reporting of data pertinent to a particular marketing problem facing the organization," according to Philip Kotler.

"Market research, according to the American Marketing Association, is the task of gathering, organizing, evaluating, and interpreting the needed marketing information."

Objectives of market research:

- To comprehend why consumers, purchase a product.
- Predict the likely amount of future sales or anticipated market share.
- Evaluate the strategies and competitive strength.
- Assess the success of existing marketing initiatives.
- Determine how satisfied customers are with a company's goods and services.

The benefits of marketing research include:

1. **Current market trends are indicated**: Marketing research keeps a company unit up to date on the most recent market developments and provides advice for addressing market circumstances with assurance. It makes production easier to accommodate consumer demands and preferences.
2. **Identifies shortcomings in marketing practices:** Medical sales people identify problems with the products, pricing, advertising, etc. It provides accurate direction with reference to several marketing-related topics. Product creation, branding, packaging, and advertising are among them.
3. **Explains customer resistance:** Customer resistance to the company's products is discovered by medical representatives. The researcher also suggests appropriate corrective actions to address the problem. This makes the goods appealing to customers.
4. **Suggests sales promotion techniques**: Marketing research enables a producer to implement appropriate sales promotion tactics, select the most practical distribution channel, set an appropriate price policy for the products, and grant discounts and concessions to dealers. It facilitates sales promotion.

5. **Guidance to marketing executives**: When formulating marketing policies, marketing executives might get advice and information from marketing research. Continuous research gives a business the confidence to take on challenging marketing situations. It serves as protection against potential market environment changes.
6. **Selection and training of sales force**: Marketing research is useful for hiring and educating workers in the sales department. It suggests incentives to be offered to staff members who work in marketing.
7. **Facilitates business expansion:** A business unit can develop and broaden its operations with the help of marketing research. Through consumer-focused marketing policies and programmer, it helps a business unit build goodwill in the market and achieve high profits.
8. **Facilitates appraisal of marketing policies**: Marketing leaders might evaluate their current marketing strategies through research activities in order to combat research findings. In accordance with the comments presented, appropriate changes to the policies are also conceivable.
9. **Suggest marketing opportunities:** New marketing opportunities and ways to effectively utilize them are suggested by marketing research. It pinpoints opportunities in both the current and future markets.
10. **Facilitates inventory study**: Utilizing marketing research, a firm can evaluate its inventory management practices and introduce more productive ways to handle inventories, including finished goods and raw materials.
11. **Provides marketing information:** Market research can provide details on various aspects of marketing. It identifies the company's relative advantages and weaknesses. Such information makes it simple for marketing executives to create future policies and offer advice, knowledge, and other approaches to marketing problems.
12. **Suggests distribution channels**: The effectiveness of current distribution channels and the necessity of making necessary adjustments to the distribution system can both be studied using marketing research.
13. **Creates progressive outlook**: Marketing research promotes an optimistic and forward-thinking attitude throughout the entire corporate structure. It encourages methodical thinking and a sense of professionalism inside the organization. Additionally, it inspires passion in marketing-related professionals. This benefits the entire company unit, bringing prosperity and stability.
14. **Social significance:** From a social perspective, marketing research is of utmost importance. It serves as a tool to make the ultimate consumer the literal ruler of the market.

Disadvantages of marketing research are:

1. **Offers suggestions and not decisions:** Marketing analysis cannot take the place of decision-making. The researcher does not offer pre-made solutions to marketing issues. Marketing issues are not directly resolved by marketing research. Only the management's process for making decisions and solving problems is aided.
2. **Fails to predict accurately:** In marketing research, attempts are made to foresee potential future circumstances. A few research investigations are conducted for this. The forecasts made, nevertheless, might not be exact. Since the future is never guaranteed, precise future predictions cannot be made through marketing research.
3. **Time-consuming activity:** The lengthier time required to complete the study and the potential for outdated results make market research a time-consuming undertaking. Even obtained data quickly becomes outdated due to the rapidly shifting market environment.
4. **Costly/Expensive activity:** Because expertise is needed for research projects, market research is an expensive endeavor. The research crew must also have advanced training in sociology, economics, and other related fields. It costs money to even hire an advertising agency or management consultant to conduct research.
5. **Uncertainty of conclusions:** The focus of marketing research is the consumer. However, it might be challenging to effectively and precisely assess consumers' purchasing intentions. This casts some doubt on the findings made based on the market study.

The following are characteristics of marketing research:

I. **Systematic and continuous activity/process**: It is a continuous process and expected given that the marketing of goods and services may occasionally bring up fresh marketing challenges. One type of research is insufficient for all marketing-related challenges. To address new difficulties and challenges in marketing, new research activities will be necessary.

II. **Wide and comprehensive in scope**: Its scope is wide since it incorporates all facets of product and service marketing. Just a few of the topics covered by market research include the introduction of new products, the identification of potential markets, the choice of suitable marketing tactics, the study of market competition and buyer interests, the initiation of a suitable marketing strategy, and sales promotion measures.

III. **Emphasizes on accurate data collection and critical analysis**: Appropriate data should be accurately and objectively gathered for marketing research. The data gathered must be trustworthy. It needs to be analysed methodically. This will give a complete view of the problem and potential solutions.

IV. **Offers benefits to the company and consumers**: The sponsoring company benefits from marketing research. It increases the company's revenue and earnings. Additionally, it improves the market's reputation and competitive capabilities. It helps a business to implement customer-focused marketing strategies. Due to marketing research initiatives, consumers also receive pleasant goods and experience more contentment.

V. **Commercial equivalent of military intelligence**: It falls within the category of commercial intelligence work. It makes planned marketing-related activities easier. It is comparable to military intelligence, wherein the environment is carefully examined before any military action is made. An instrument for managerial intelligence is marketing analysis.

VI. **Tool for managerial decisions**: It acts as a tool in management's hands for identifying, analysing, and resolving marketing problems. It is a tool for making decisions. It offers potential solutions for managers to think about and choose. Marketing analysis is a supplement to judgement, never its replacement.

VII. **Applied research**: It is applied knowledge that addresses a specific marketing issue and offers potential solutions along with their potential results.

VIII. **Reduces the gap between the producers and consumers**: A medical representative is a crucial addition to aggressive marketing. Understanding consumer requirements and expectations is helped by this. It narrows the gap between producers and customers and modifies marketing efforts to better meet consumer wants.

IX. **Marketing research has limitations**: Only prospective marketing research solutions are presented to the marketing manager for review and selection.

X. **Use of different methods**: Various techniques can be used to do market research. Data can be gathered through surveys or other techniques. The method that is best for carrying out the research project must be chosen by the researcher. This decision is crucial since the research methodology utilised determines how well the research will turn out.

Role of market research in marketing:

Market research concentrated on the actions that need to be taken, as shown in **Figure 1.8**, and the specific roles that need to be played, as shown below:

(i) **Framing and implementing product policies**: Problems in the creation and execution of product policies are efficiently resolved through marketing research. These product policies include product development, product line selection, input purchases, inventories, plant location and layout, production planning and control, plant maintenance, waste management, cost and quality control, finances, and personnel.

(ii) **Designing and executing marketing strategies**: The dynamic, active, and widespread plan for achieving corporate objectives is marketing strategy. A manufacturer who achieves the best possible balancing of controllable and uncontrollable aspects is successful. Because of the current market conditions, firms must be customer-focused. As a result, all of their efforts are in line with consumer wants, moods, and budgets. Because they are inside to the company, a manufacturer can control elements like product price, promotion, and distribution.

(iii) **Location of the outlet:** The location of a distributor's selling outlet is critical to his success. The region and the particular site are given top attention while choosing the location. Finding places with a dense and wealthy population, a supportive corporate environment, and appropriate transportation infrastructure is made easier with the use of marketing research. When choosing a location, consideration is given to factors that make the area convenient, such as access to clients, banking services, and rival businesses.

(iv) **Shaping and improving store image:** The store stands out from its rivals thanks to its unique personality. This needs to be highlighted in order to increase public trust in the company's products and services. An individual's perception of a business is the result of his views toward numerous facets of that business. These attitudes impact perception, motivation, interpersonal interaction, characteristics, and self-concept in turn. Consumer perceptions regarding stores are influenced by elements such as buying habits and store attributes, according to market research studies.

(v) **Distribution Cost Control and Reduction:** We have received things from the current industrial system that are better quality, less expensive, and insufficiently plentiful. The price of distribution has, however, risen faster than the price of manufacturing. Marketing research can be used to control the prices of transportation, warehousing, finance, advertising, consumer services, credit returns, and other modifications.

(vi) **In Advertising**: In this area marketing research plays an important role, if budget appropriations for advertisement is achieved. In preparation and placement of advertisements, effective measurements of advertising are taken which will analyse and improve the of image of corporate brand and stores.

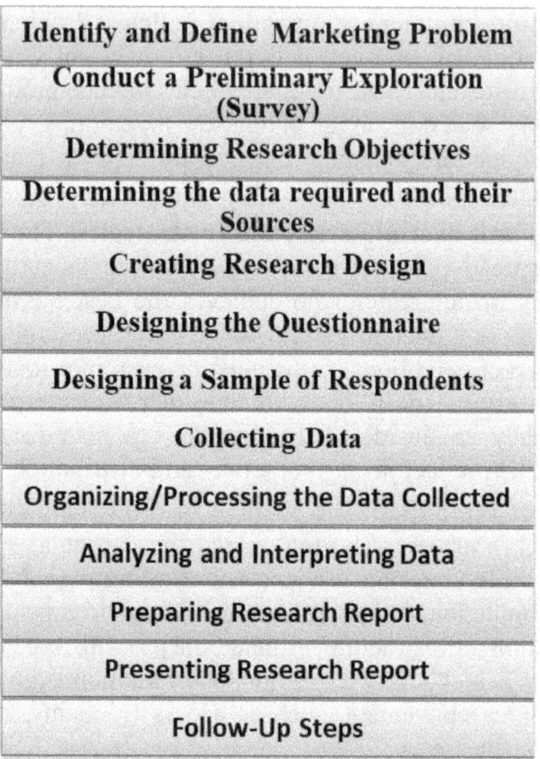

Figure 1.8 Steps for market research

1.2.9.1 Areas of Market Research

The market information will be provided by the market research done in various areas as:

Marketing mix research: It is done by analyzing the research of product, price, distribution and promotion of a product.

(i) **Product research** is carried to study the acceptance of new product in market as it involves the study of product quality, features and design. The product actual use is also studied with new method of use as well as its packaging and packaging design. After this testing of a product is done using methods:

Test marketing: Means introducing a new product in one or two target markets and evaluating the response in those markets.

Concept testing: Testing the consumer reaction to the description of a product concept rather than the actual product.

Testing the packaging concept: New concepts related to packaging of product are evaluated in market.

Attribute testing: Some special or modified features of product like sugar coating, special flavor of the drug is evaluated in comparison to unmodified product.

Assessment of products strength and weakness: The major attributes as well as the weak points of a product are evaluated by market research.

(ii) **Pricing Research**: This includes research related to determination of pricing strategy for product. The competitor's reaction to the price strategy of the firm and determination of price elasticity of demand means how sensitive is demand to the variation in price. This will analyze the pricing strategy of customers.

(iii) **Distribution research**: It involves the analysis of effectiveness of different distribution channels or intermediates by determining of competitor's strategy with reference to distribution margins. The distribution results in identification of least cost location of warehouse.

(iv) **Promotion research**: This refers to the analysis and evaluation of different elements of promotion mix i.e. advertising, sales promotion, personal selling and publicity and their impact on sales volume of product.

Market research: This research determines the market specially for gathering information on numbers of competitors in market. Market share of different competitors will show the growth rate of market. Latest trends and development in market give the indications of potential of new products launched in market.

Consumer research: This research includes gathering the information on consumers buying behavior and patters like how consumers buy, where they buy, when and why they buy a product.

Competition Research: It is performed to reveal the competitive position of company as well as to know the strengths and weakness. It includes- Study of competitor's products, measuring the impact of competitor's price, advertising channels and sales method.

Market measurement research: This will be the togetherness of demand, market performance and motivation.

(i) **Demand research**: Is carried out to find, how much a particular product can be sold in a market. It includes- Determination of market potential, short run and long run sales potential.

(ii) **Market performance research**: It is carried on to measure the existing market. It includes- study of market size, market profits, market segments and sales forecasting.

(iii) **Motivation research**: It studies the buyer behavior and attitude so as to expand the market in a specific place. It includes, the study of consumer profile, study of consumer taste and preferences, study of consumer dissatisfaction and sources of dissatisfaction.

1.2.9.1.1 Market Research Process

There are five general steps to be followed for any market research.

(i) ***Preliminary investigation***: The first step of market research determines and analyze the marketing problems to be solved, along with this immediate objectives and ultimate goal of company are carefully stated.

(ii) ***Research design***: This is the master plan for conducting the market research. It gives specification of methods and procedures for acquiring the information needed for solving the problems.

(iii) ***Data collection***: Most critical step in market research. Data may be collected by two sources-

Primary data: It involves the information obtained from original sources by researcher. The important source is consumers (patients), doctors, pharmacist, nurses etc. This data is valuable because it gives latest and very accurate information about current market trends but this method of data collection is costly and time taking. The method for obtaining primary data includes:

(a) **Mail/Questionnaire survey**: It is very effective method of survey in which printed copy of questionnaire along with the covering letter is mailed to the survey respondents. A reply letter (stamped) is also sent so that respondent needs not to pay. This is significant due to reasons of no geographic limitations and secure the information from respondents of distant areas. Sometimes possibility is that many persons don't respond and don't send reply letter back. This method has low degree of accuracy because many people give unreal or rubbish information in letters.

(b) **Personal interview** (field survey) It is the direct form of investigation involving face to face communication with people. It is more flexible form of data collection. It is an accurate method with low degree of refusal.

(c) **Observation Approach**: In this method obvious behavior of consumers is observed by trained observers without asking any question to consumers. Example, Observing the response of consumers in a chemist shop or checking the packs of medicines, or rejections, present in the chemist shop etc.

(d) **Online survey**: It involves collection of information through internet. The company can put a questionnaire on its website and offer incentive to people on answering the question. It is an inexpensive method and response are obtained very fast but have demerits that only few people access to internet which show very unreliable data collection because many persons just give wrong answers for fun when nobody is seeing them.

Secondary Data: Here data reveals the information that is in published/ semi published form Example of published market surveys, government publications and reports, general library research sources, All advertising media specially newspaper, magazines, pharmaceutical journals etc., drug association and other technical healthcare organization, universities publications (articles and thesis). This method is quicker and economical for data collection. Provides the data for all subjects and areas and useful when primary data is difficult to obtain or giving variable results. Only disadvantage is that it is not so reliable, biased, lacks reliability.

(iv) *Data Analysis*: Once the data is collected, it is processed, analyzed and interpreted. The steps involved in data analysis are:

- **Data tabulation**: All tabulation of data is done by sorting it into different categories and counting the number of cases that belong to each category.
- **Coding of data**: After tabulation the individual response categories are coded by certain numbers.
- **Frequency table**: A frequency table is prepared that shows the counts or occurrence of individual categories among the sample.
- **Data interpretation**: The data is interpreted by using advanced statistical techniques like t -test, chi- square test, binominal test, multivariate test etc.

(v) ***Report preparation***: It is an important stage of marketing research. The report is prepared for effectively communicating the research information. The essential components of research report format are:

- **Tile page**: It is required for report identification. It includes subject of report, for whom it is prepared and by whom it is prepared.
- **Contents (Index)**: An index is an outline of how the study report is organized. It divides the whole report into several parts (sections, subsections) that help in locating the different components of report.
- **Introduction**: It serves to present the background of research report and highlights the major issues which are examined in report.
- **Statement of purpose**: A statement of research objectives clarifies the issues covered in the study.
- **Research methodology**: It includes how the research study has been performed and includes the method of data collection, sources used for data collection, sample size, analysis and procedure.
- **Data analysis section**: It includes tables, graphs, and statistical analysis tools.
- **Results**: These are the basic outcomes of the study. The result involves textual comments on the interpreted findings of research.

- **Conclusions and recommendations**: Based on the research results a conclusion paragraphs are prepared, which shows researchers opinions on the issues examined after conclusion. Future prospects of the study are provided.
- **Appendix**: The supporting documents like questionnaires and bibliography are provided at the end of report.

Answer the Following Questions

Q1. Define marketing. Discuss the features and scope of marketing.

Q2. What is the procedure for marketing?

Q3. Differentiate between buying and selling.

Q4. What are the skills required by selling agent working in an organization?

Q5. Explain the sources of getting feedback information in marketing department and mention it's significance.

Q6. Differentiate between selling and marketing.

Q7. Discuss the microenvironment and microenvironment in marketing.

Q8. Write the steps involved in consumer's decision for accepting or buying new products coming in market involves.

Q9. Differentiate between quantitative and qualitative aspects of pharmaceutical market.

Q10. Write a note on career opportunity in marketing.

Q11. Explain the Role of market research in marketing

Q12. Define market research. Elaborate various characteristics of marketing research.

Q.13. Explain various methods of forecasting.

Q.14. Give an account on Patient's choice of physician.

Q.15. Discuss various strategies that a company may adopt for market targeting.

Q.16. Write note on prescribing habits of the physician.

Q.17. Give an exhaustive detailed outline on Market segmentation.

Q.18. Write the note on size and composition of the market.

Q.19. Discuss various concepts of marketing.

Q.20. Define selling. Explain steps involved in selling process.

MCQs

1. Marketing is a process which aims;
   - A. Production
   - B. Profit making
   - C. Selling products
   - D. Satisfaction of customer needs
2. Which forms of human needs that take as shaped by culture and individual personality.
   - A. Want
   - B. Demand
   - C. Need
   - D. Social Need
3. Marketer often used which of the term to cover various grouping of customers.
   - A. Buying power
   - B. Demographic segment
   - C. Markct
   - D. People
4. ------------------------- buy products and ---------------- use product.
   - A. Consumer and customers
   - B. Customers and consumer
   - C. Buyers and sellers
   - D. Buyers and customers
5. A ---------------------is a detailed version of the idea stated in a meaningful consumer terms.
   - A. Product concept
   - B. Product feature
   - C. Product idea
   - D. Product image
6. Our marketing mechanism is deemed successful only if
   - A. We get money from him
   - B. Customer is fully satisfied by our products and services
   - C. We can sell more than our competitors
   - D. We can make more profit than our competitors
7. Marketing is a process that creates, communicates and delivers.
   - A. Services to customers
   - B. Products to customers
   - C. Value to customers
   - D. Materialistic benefits to the customer
8. Technology environment is part of firms.
   - A. Microenvironment
   - B. Macro environment
   - C. Task environment
   - D. Depends upon the nature of the firm's product range

9. If we try to meet more and more people and share our emotions with them, we are satisfying the following need.
   A. Self-actualization B. Physiological
   C. Security D. Social
10. If a consumer understands the right message in a wrong way, the firm would stand to
   A. Gain B. Remain unaffected
   C. Lose heavily D. None of those
11. The term market is derived from which word.
   A. Latin B. Greek
   C. German D. None
12. Testing of new products in market is done by
   A. Test marketing B. Concept testing
   C. Attribute testing D. All
13. Among the following which one is not a part of Data analysis?
   A. Data tabulation B. Data coding
   C. Data interpretation D. Data screening
14. One product all segments are the strategy of
   A. Differentiated marketing B. Concentrated marketing
   C. Vertical marketing D. Undifferentiated marketing
15. Several products- Several segments is the principle of;
   A. Differentiated marketing B. Concentrated marketing
   C. Vertical marketing D. Undifferentiated marketing
16. AMA is elaborated as;
   A. American marketing association
   B. American marketing assets
   C. American management association
   D. American management assets
17. The term marketing mix was introduced by one of the following;
   A. Borden B. Kotlers
   C. Smith D. Taylor
18. When segmentation is based on age, gender, income etc. then it is called as
   A. Demographic B. Geographic
   C. Psychographic D. Behavioral

19. Hard core loyal means
    A. Loyal to one brand
    B. Loyal to two or three brands
    C. Psychographic
    D. No loyalty
20. Which one is odd among the following with respect to marketing?
    A. Product concept
    B. Marketing concept
    C. Selling concept
    D. Pricing concept
21. Point out the exact sequence of selling process
    A. Prospecting-Approaching-Presentation
    B. Prospecting- -Presentation-Approaching
    C. Approaching -Prospecting -Presentation
    D. Presentation -Prospecting-Approaching
22. Porter's analysis is based on
    A. Five forces
    B. Six forces
    C. Seven forces
    D. Four forces
23. Hierarchy pyramid is the concept of
    A. Maslow's
    B. Kotler
    C. Borden
    D. Smith
24. Self-respect and prestige comes under
    A. Safety need
    B. Social need
    C. Esteem need
    D. Self-actualization need
25. Highly traditional and price conscious consumers are called as
    A. Early adopters
    B. Late majority
    C. Innovators
    D. Laggards
26. When no of sellers (company) is very few in the market, it is called
    A. Monopolistic market
    B. Monopsonistic
    C. Bilateral market
    D. None
27. The meaning of Value is understood as
    A. Benefit / Cost
    B. Cost/Benefit
    C. Cost-Benefit
    D. None
28. Competitors, Suppliers and Customers comes under the
    A. Internal environment
    B. Micro environment
    C. Macro environment
    D. All

29. How many numbers of competitive forces jointly determine the strength of industry competition and profitability.

    A. 5
    B. 6
    C. 7
    D. 8

30. Consumer behavior is the process whereby individuals decide what, when, where, how and from whom to purchase goods and services given by

    A. Walters and Paul
    B. Smith
    C. Robinson
    D. Taylor

Answer key to MCQs

1. D, 2. D, 3. C, 4. B, 5. A, 6. B, 7. C, 8. B, 9. D, 10.C,
11. A, 12. D, 13. D, 14. D, 15. A, 16. A, 17. A, 18. A, 19. A, 20.D,
21. A, 22. A, 23. A, 24. C, 25. D, 26. A, 27. A, 28. B, 29. D, 30. A

CHAPTER 2

Product Decision

2.1 Product Decision

Any product in the market comes with the vision of the prospective of product, market etc.

Products: Whatever a marketer can receive to satiate a want or demand. Physical items are among the products that are marketed (Tablets, capsules etc.), Services (Pharmaceutical consultancy, finance companies etc.), Person (Virat Kohli, Amitabh Bachan etc.), Places (Goa, Delhi, Mumbai etc.), Organization (Indian medical associations, Indian Pharmaceutical Associations etc.) and ideas (Family planning etc.).

2.1.1 Product Classification

Traditionally, objects were divided into three categories based on their qualities of sturdiness, tangibility, and usage (whether industrial or consumer):

A. ***Non-durable product***: These are the tangible products that are often used just once or twice, such as oil, soap, toothpaste, salt, and other such things. The best strategy is to make these commodities broadly accessible, price them at a low markup, and sell them widely to promote trial and foster preference because they are quickly consumed and bought frequently. These products need significant advertising, have a low profit margin, and have to be easily accessible.

B. ***Durable products***: Products with a longer lifespan and number of uses. These products offer a large profit margin, call for greater in-person sales efforts, demand after-sale services, etc. Example, refrigerators, machine tools, and clothing.

C. ***Services***: These are perishable, variable, inseparable, and intangible goods. They typically need higher quality control, supplier credibility, and agility as a result. Examples include haircuts, legal advice, and pharma industry consultants, repairing of highly sophisticated machine.

In the current era the product classification is as follows:

A. ***Consumer products:*** The end consumer purchases these things for his own use. The big market has to satisfy various consumer wants or needs of products bought by customers for final consumption. As a result, businesses should produce a variety of goods. Consequently, consumer goods are categorized as:

(i) **Convenience products:** Convenience goods are typically inexpensive, readily accessible items that customers frequently purchase without prior preparation or research, and with the least amount of comparison shopping and purchasing effort. These commodities are available for purchase by consumers through a variety of distribution channels, including every retail location. This category includes food products like rice, wheat flour, salt, sugar, milk, and other fast-moving consumer goods (FMCG). Its feature includes that they are easily available with minimum time and effort. These include necessities since there is a consistent and low demand for them. Since there is a large surplus of these commodities compared to consumer demand, there is fierce rivalry for them. Sales promotion programs including discounts, freebies, rebates, and others aid in the marketing of these goods.

(ii) **Shopping products**: These items take a lot of time and effort to produce. A consumer compares the quality, price, style, etc. at various stores before planning. These goods are sold through a few carefully chosen distribution channels. Television, air conditioners, automobiles, furniture, hotel and airline services, and tourism services are some examples. The main features of these products includes durability, good profit margin, comparable market for consumers. A consumer can plan the purchase of these products which can play an important role if the product is purchased from retailer.

(iii) **Speciality products**: Products with unique qualities or brand recognition that make consumers willing to invest a significant amount of time and effort in their purchase. It includes main features of specialty products as high cost, availability at few places. Since such products are relatively less available hence an aggressive promotion is essential for the sale of such products.

(iv) **Unsought products**: These goods are said to as unsought because consumers frequently either unaware of them or are aware of them but do not typically consider purchasing them. Marketing professionals need intensive promotional efforts to promote their goods.

B. ***Industrial products*****:** Products purchased by people or organizations for use in running a business or further processing. The following products are necessary for industry:

- **Materials and parts include:** Examples of raw materials include agricultural products, chemicals used in the pharmaceutical sector, crude oil, iron ore, manufactured items like iron, cement, and wires, and component parts like small motors, tyres, and castings.

- **Capital items:** Capital goods, which include installations like factories and offices as well as enduring equipment like generators, computer systems, and elevators, as well as ancillary equipment like tools and office supplies, aid in production or operation.
- **Services:** Services include upkeep and repair work including machine and computer maintenance and repair work, legal and consulting services, and advertising and marketing services.

C. ***Pharmaceutical products*:** They come under consumer products as they are brought by final consumers or patients for their consumption or their family members may consume. Pharmaceutical products are known as medicine or drugs, they are important apart of traditional or modern system of medicines. So, and these products should be of good quality, effective and safe. The pharmaceutical products are very well identified by ***prescription drugs*** (Allopathic, Homeopathic, Ayurvedic and Unani medicines) and by ***non- prescription drugs*** (as Over the counter drug).

There are five different levels of products:

The core benefit is the most basic level: The fundamental level is where the utility offers the product, creates the core product, or renders the core service. For instance, the primary product of the pharmaceutical sector is drugs, whereas the primary product of a hospital is patient care. The less participants there are in the market and the less competition, the more distinctive the core product is engineering.

Second level is the basic products: The basic product is the second level as seen in cases of hospitals it is the building, specialty doctors, clinical pharmacist, drug information center, laboratories etc. This is the basic version of the products like the primary drug constituent and hospitals having doctors, nurses, beds, laboratories and its own pharmacy.

The desired outcome is at the third level: Based on facilities and convenience providers like Apollo hospital, Max hospital, AIIMS Hospital, and services for critical patients, expected from these, there is a set of characteristics and conditions that buyers typically expect and agree to purchase for the products. As the brand grows in reputation, the expectations of the consumer will be grown in all aspects. If it fails to deliver on this anticipated product, it will also have an impact on the fundamental product.

Fourth level is the augmented products: That mean the customers desires beyond their expectations like relief from the disease within the prescribed time frame, Telemedicine facilities, specialized patient care, portable dialysis aid, hearing aids. Max and Apollo hospital giving delightful patient care, transplantation of heart, kidney, liver, 24 hrs. ambulance services etc. The customer's desire, which can be made into reality, is the augmented product.

Fifth level is the potential product: Every business examines the potential of the products it already sells, considering all the upgrades and changes that the product may eventually experience in the future. Examples include drone-delivered medical supplies, a stem-cell treatment for diabetes, a wristband that can read your thoughts, an ultrasound in your pocket, artificial intelligence that can diagnose cancer, etc. Every item on the market has potential. It is the prospective product with the possibility to create a millionaire among the five product levels.

2.2 Product Line and Product Mix Decisions

2.2.1 Product Line

A collection of closely related products from one product class that employ the same marketing strategy to target the same consumer demographics. *Proctor and Gamble* which began business as a soap manufacturer has developed or acquired product lines in among other areas like foods, coffee, paper products and baby products.

According to **Philip Kotler**, a product line is a collection of goods that are functionally similar or otherwise closely related. advertised through the same channel or outlets, sold to the same target groups, and priced within specified ranges.

Examples of Product line as:

- **Amul** provides a range of items that are all closely related, including milk, butter, ghee, dahi, yoghurt, ice cream, Srikhand, Gulab jamun, flavor-infused milk, chocolate, etc.
- **Nike** is a global corporation that markets a variety of sports footwear, apparel, and equipment product lines.
- **Nestle** are: Nespresso, Nescafe, Milo, Illuma, NAN, Garden of Life, Purina ONE, Felix, Merrick, Nido, Haagen Dazs, Maggi, Kit Kat, Pure Life.
- **Coca-Cola** are: Coca-Cola, Sprite, Fanta, Dasani, Smart water, Minute Maid, Innocent, Fresca, Aquaris, Fair life, Honest, Georgia Coffee, Costa Coffee.

Multiple product line: The likelihood of pleasing more consumers and increasing productivity grows with these having a large selection of product lines. It helps to reduce the risk associated with being more dependent on one product and market segment.

Example **Johnson and Johnson** had introduced multiple product line in wound care sections-

- Band Aid which is first medicated dressing.
- Savlon, antiseptic lotion which heals without hurting the skin.

- Johnsonplast, is a surgical dressing product for boils, burns and large wounds that cannot be covered by Band Aid.
- Johnsons Capsicum plaster is used for relieving muscular pains, backaches, swollen joints and boils.

2.2.1.1 Product Line Extension

"Product line extension" is the practice of a company introducing new products with the same brand name but distinct attributes such as tastes, forms, colors, additional ingredients, various package sizes, etc. An established product is modified through a line extension. A new change of an existing molecular entity, such as switching from immediate release to prolonged release, or a new formulation of an existing product can both constitute a variant. Numerous line extension varieties, such as innovative versus older line extensions and non-branded versus branded, are described in general marketing literature. There examples are as mentioned below:

- Colgate with hemp seed oil
- Johnson and Johnson in their feminine products category launched Stayfree in various variants like napkins, silky dry and tampoons. They added an extension to Stayfree-New Stayfree Secure.
- Head and Shoulders Menthol and Pantene Pro-V anti-dandruff were introduced by Proctor and Gamble.
- Dabur India Ltd expanded the Vatika brand to include a conditioning shampoo called Vatika Henna Cream.
- Iodex developed Iodex Power Cream to address the lower waist back issue. For those with asporty outlook, the Iodex sport was introduced.
- Karnataka Soap and detergent Ltd has launched Mysore Sandal Gold, a sandal wood oil soap to be sold in India.

2.2.1.2 Product Line Stretching

Product line stretching is the practice of adding additional products to an existing product line. This suggests that the company makes a greater variety of products, from high-end to low-end goods. There are three methods for expanding the product line, including:

Stretching downwards: When a company launches lower-quality items to reach the upper middle class, as in the case of "Tata Motors," a "downward stretch of the product line" occurs. Now it has launched "Tata Nano" to target the lower middle class also. This is an example of downward stretching

For a variety of reasons, businesses would downward expand their product line:

- By stifling rival activity and rival product offerings
- To compete in the market's lower price range, especially if it is a high-volume segment.

- extend the positioning of their brand so that it is perceived as a more accessible one generally.

When a company launches lower-quality items to reach the upper middle class, as in the case of "Tata Motors," a "downward stretch of the product line" occurs.

***Stretching upwards*:** The opposite of a product line stretching downward is one stretching upward. Here, the business introduces products that are superior to its standard offerings in terms of quality. For instance, Maruti Suzuki Limited introduced the "Maruti 800" automobile in 1983. Later in 1985, it introduced "Maruti Gypsy," a more expensive brand than "Maruti 800." This is an ascent stretch.

Companies frequently aim for high-quality items and expand their product line due to a number of factors, including:

- Higher unit margins are frequently found in high-quality products, which, with a low turnover, can be very profitable.
- Creating a prestige brand for the entire company by offering high-quality products often enables price premiums to be applied to the entire product line.

An upward stretch has disadvantages, including:

- The development of new brand names may be necessary if the current brand equity and image don't transcend to the high-end of the market.
- Rivals who now have positions in the top end of the market might try to hang onto them.
- Additional or expanded distribution channels may be required to support high-quality products.

The level of sales volume could not be sufficient because there is normally less turnover at this time of year in the market.

Stretching both upward and downward simultaneously:

Sometimes businesses will try to spread their product lines both upward and lower, however this is uncommon and more probable in emerging or growing markets.

This would happen when companies work to provide a wide range of product options while simultaneously rapidly expanding their product line to reduce the motivation for new competitors. With this product line strategy, the company seeks to control or dominate the market.

2.2.1.3 Product Line Filling

This strategy for product line extensions is more prevalent since it enables additional products to be offered that are consistent with the brand's initial positioning. Therefore, rather of expanding into the higher or lower quality end of the market, the brand simply introduces additional choices. As an illustration, Maruti Suzuki introduced the Alto in 2000 as a product that fell between the Maruti 800 and Maruti Zen models. As a result, the company has tried to close the price difference between the "Maruti 800" and the "Maruti Zen." This means that "Alto" costs somewhere between "Maruti 800" and "Maruti Zen.".

Product Line Pruning: Removing underperforming products from a product line is what it refers to. As an illustration, "Toyota Kirloskar" discontinued the popular brand "Qualis" because it didn't think it was profitable enough.

2.2.2 Product Mix

The product mix refers to all of the product lines that a seller provides to its clients on the market. There may be a variety of items in each of the company's product lines. The company's product mix is the concoction of numerous product lines. It entails organizing, creating, and creating the appropriate kind of goods and services that the organization will sell. Branding, packaging, trademarking, labelling, etc. are all included. The product mix of a corporation refers to the assortment of goods it sells.

Each product mix has a depth and is calculated on the basis of models, colors, sizes available in each product line. If a company possesses, say antibiotics product line. The dosage forms, such as tablets, capsules, vaginal suppositories, injections, ear drops, and syrups in, say, a few package sizes, will aid in determining the depth of the product mix.

The specific advantage of the varied product mix is that whenever there is a loss in one product line, the company will be able to offset its loss by the profit it makes in other divisions or product line.

For example;

(i) AAB Pharmaceutical gives a wide range of pharmaceutical product mix as:

- **Anti-infectives**: Amoxicillin, Ampicillin, Chloramphenicol, Co-trimoxazole, Erythromycin, Gentamycin, Penicillin G, Tetracycline, Penicillin VK
- **Anti-cold**: Bromhexine, Dextromethorphan, Guaifenesin
- **Analgesic- Antipyretic**: Paracetamol, Diclofenac sodium 50 mg, Mefenamic acid 500 mg cap,
- **Vitamins**: Minerals-Ascorbic acid, Thiamine, Ferrous fumarate tab.
- **Anti-TB:** Pyrazinamide, Rifampicin, Isoniazid, Ethambutol,
- **Steroidal:** Antihistaminic-Chlorphenamine maleate, Dexamethasone, Prednisolone.

(ii) Product Mix of Wipro

- **FMCG**: Soap, Food, Deodorants, Baby Dippers, Shampoo, Agarbatties.
- **Computer solutions**: Laptop, desktop, Printers, CPU, UPS, Mouse.
- **Lighting and Transport**: Street light, dumpers, Torches, CFL and battery.

(iii) Products of Hindustan Lever Ltd, HLL offers a variety of product categories, such as soaps, detergents, and toothpaste, among others.

(iv) Dabur India Ltd started as an ayurvedic drugs company and now has a range of products from consumer goods to commodities.

(v) Johnson and Johnson Ltd the multinational giant is a classic example. With more than 100 different items available for purchase in the consumer, pharmaceutical, and professional markets, it is the most complete maker and marketer of health care products.

2.2.2.1 Dimensions of Product Mix

This has been classified by the company's own strategy:

Product mix width: how many different product lines in total the organization offers to customers. Example: The product mix from Jyothy laboratories comprises six lines. Hence, width mention in **Table 2.1** for the particular company is six in number.

Table 2.1 Product mix dimension for a company

Fabric care	Family insecticides	Dishwasher cleaners	Fragrances	Individual care	Allied Business
Ujala washing powder, Ujala supreme Stiff and Shine.	Maxo cyclothrin coil, Max vaporizer, Max aerosol.	Exo dish wash liquid, Exo dish wash bar.	Maya (8,15,20,40 and 100 sticks).	Jeeva natural (Coconut milk with milk protein, Coconut milk with Kasturi, Coconut milk with Jasmine).	Marketing of Godrej Tea, Marketing of Ekta dhoop.

Product mix length: The samples shown below show the total number of things that company carries in its product line.

Hindustan Lever Limited

- Soap (Product line Length = 5): Lifebuoy, Lux, Dove, Breeze, Pears
- Detergent (Product line Length = 4): Wheel, Rin, Surf, Surf Excel
- Shampoo Product line Length = 2): Clinic, Sunsilk
- Toothpaste (Product line Length = 2): Close-Up, Pepsodent

Here Product mix length is 13

Jyothy laboratories Three products from the fabric care section are Ujala ultimate, Ujala laundry detergent, Stiff, and shine.

***Product mix Depth*:** How many variations of each product in the series are available. Example: Three distinct iterations of "Jyothy laboratories" Jeeva natural exist (Coconut milk with milk protein, Coconut milk with Kasturi, Coconut milk with Jasmine in 75 gm packs).

2.3 Product Life Cycle (PLC)

It describes the products life history starting from its birth to old age and death. Plotting the sales volume of a product over time will allow you to graph a PLC. The curve that results is typically bell-shaped. There are four stages:

(a) **Introductory phase**: Because the product is still in its early stages of development, fewer individuals are aware of it. As a result, the corporation uses aggressive product awareness and sales marketing tactics. Sales are now low.

Profits are often negative or at least break-even after one to two years due to poor sales outputs, low manufacturing, expensive promotional charges, and other marketing-related expenses.

(b) **Growth phase**: The product sales accelerate in market resulting in better profits to the company. Company focuses on building better distribution networks to make the product widespread. The number of competitors also grows.

Additionally, it has greatly increased consumer support, patronage, and volume of repeat purchases.

Strategies to improve the growth of a product:

- An innovation or continuous improvement product.
- Expanding distribution channels, breaking into new markets, new market segments, and offering trade outlets more incentives to sell and recommend the product to customers.
- More aggressive and proactive promotions efforts.

(c) **Maturity Phase**: A time frame in which the product's sales volume slows down, plateaus, or remains stable. This could be as a result of the target market segment being intensively targeted by a large competitor, or as a result of many identical products competing in the same market area.

Strategies:

- Increase product support through advertising and other campaigns.
- Budget increases for research and development.
- Reduction in prices through additional discounts, free goods offered as part of product promotions, additional compensation or assistance, and other perks to entice retailers to stock up on the item.

(d) **Decline Phase**: This time period has been marked by long-term, consistent declines in the sales and profits of most product types and brands. The tactics developed at this stage are:

- Either make major modifications in the product design, so as to influence the consumers or take back the product from market.

Example. For tuberculosis INH (Isoniazid), PAS (Para amino-salicylic acid) and streptomycin were introduced and used as first line treatment. Then new drugs came into market like Ethambutol and Rifampicin. Since new drugs are more potent, the use of older drugs decline. Hence, INH manufacturer's users extended it life cycle by combination with Rifampicin. Thus this synergistic combination of INH + Rifampicin has captured 90% of anti-tubercular market.

- Reduce the cost.
- Cease activities in marginal trade channels and smaller market segments.
- Offer the pharmaceutical product for sale to another business or simply liquidate it at a loss.

2.4 Product Portfolio Analysis

Examining a product portfolio means looking at the range of products a business offers. Making sure that the business has successful, high-performing products as well as new products in development to replace current products once they reach the decline stage of the product life cycle is its aim.

The collection of all a company's goods or services is called a **product portfolio**. A product portfolio is a grouping of the goods and services that a business provides to its target market. It includes every product offered, from those that were first introduced and made available when the brand was first established to those that are currently available and those that are in the works.

Pharmaceutical companies like Sun Pharma offer a variety of products, such as generics, branded generics, specialised, difficult-to-make, technology-intensive drugs, over-the-counter (OTC), anti-retrovirals (ARVs), active pharmaceutical ingredients (APIs), and intermediates.

Product Portfolio Management:

1. Product portfolio management is one of the most crucial elements of the entire business strategy since it helps the organisation achieve its overall business goals and plans for future product lines in accordance with those goals.
2. It is a vital instrument for both corporate financial planning for the business and for investors conducting equities research and calculating return on investments.

3. The management of the business can learn a lot from a comprehensive analysis of the product portfolio, which includes stock type, brand growth prospects, high-margin products, income contributions from each and every product offered to the market, market share of each product, operational risk, and market leadership.
4. Too many projects are frequently in process. The main goal of product portfolio management is to determine which initiatives are effectively linked with the overall business strategy and objectives.

Business importance of a product portfolio:

1. **Product innovation:** It is essential to stick to the strategy of having a product portfolio and routinely assessing it in order to plan and build a new and innovative line of items to be offered to the target market. It is helpful to understand the types and nature of goods that are valued and preferred by customers in order to introduce a new line of products that are not only imaginative and novel in their conception but also suit the tastes and preferences of the target market.
2. **Tax benefits:** In order to structure investments and all other financial parts of the firm, regular management and analysis of the Product Portfolio is helpful. This results in a number of tax benefits.
3. **Aligns projects with the business' strategy:** The product offerings and the revenue streams that underpin them must reflect the company's long-term objectives and business strategy. Because only then will the company be able to accomplish its aims and targets, which include increased sales, increased profitability, a competitive advantage, and an increased market share. A well-managed product portfolio makes it easier for management to align ongoing projects with the company's overall business strategy and objective.
4. **Visualize the entire products-line:** Drawing comparisons by looking at and examining the operations, revenue generation, and other characteristics of each and every product the firm offers separately can be very time-consuming and ineffective. But now that the Product Portfolio is in place, all of the management's key figures may view the complete portfolio of all previous, current, and forthcoming things, which has a wider spectrum.
5. **Effective allocation of resources:** Possessing and managing the product portfolio helps the business allocate its many resources, such as cash, labour, and manufacturing space, more effectively. It helps in determining which products are acting as cash cows for firms, which have the potential to take up a greater market share but require managerial assistance, and which are superfluous and should be taken off the market.
6. **Data for the key members of the management:** It helps by providing the key members of management with the necessary and crucial information that tells them about the effectiveness of the products in the market, income

generated by each product, market share, client preferences, and the need for any type of tweaking or innovation in any products, among other things, that assists in the design and implementation of the upcoming business plans and strategies.

7. **Cash flow:** The business needs a steady stream of income for day-to-day expenses like paying overhead, employee wages, and other costs, as well as for investments in the current and upcoming product lines. The company's cash flow problems are resolved with proper portfolio planning and management since it identifies the goods that generate the highest revenues and allows the company to focus the greatest amount of resources on those products.
8. **Synergy within the internal team:** The company's operations and all of its products are not handled by a single person, but rather by a number of departments and team members assembled by the management of the organization. This situation primarily applies to big corporations with a broad range of items on the market. A suitable Products Portfolio is kept in the same location where meetings and discussions between team members take place, resulting in a targeted business strategy and firm operations that work together effectively to achieve long-term goals and objectives.
9. **Proper selection of the target industry:** Management may better understand why some product lines do phenomenally well and act as the company's cash cows while others fall short of expectations and fail to meet desired objectives with the help of the products portfolio. If the latter is having a difficulty with the items not being promoted to the required target market and audience, the Products Portfolio's elements and strategies can assist in resolving this issue.

Parameters for framing Product Portfolio:

- **Amplitude**: Is a metric that accounts for the various product lines that make up the company's product portfolio.
- **Length**: This figure represents the total number of goods that the business produces and sells.
- **Depth**: This is the measurement used to depict the models in each item in the line, whether it be by size or another variation.
- **Consistency**: Examines the degree of resemblance between the lines based on consumer usage patterns, production techniques, distribution methods, price, etc.

2.4.1 Product Portfolio Planning

It is crucial in two ways:

First, by reviewing the portfolio as it stands today: Reviewing elements like is advised to improve decision-making because they include:

- Product design and potential upgrades.
- Production techniques, including ways to gauge advancement or methodological senescence.
- The level of user security provided.
- Management development is crucial since those in charge of administrative management must make crucial judgments that could affect the products' profitability.

Secondly the growth matrix of product portfolio

Boston Consulting Group (BCG) Matrix: The most popular technique used by businesses to evaluate their product lines. The Boston Consulting Group developed it in the 1970s to help businesses decide how to best allocate their limited time and financial resources.

Two pieces of information are needed for the BCG (Boston Consulting Group) matrix: the market share of each product and the rate of market expansion overall. This allows for the classification of products into the following four groups: Cash Cow, Question Mark, Star, or Dog and Problem Child.

(i) **BCG Matrix Cash Cow**- A product with **"high market shares and low market growth"** is referred to as a cash cow. High market share indicates a strong product in that market, which may allow the company to sell it for a high price. Because the market isn't increasing, it can get by with relatively little marketing spending thanks to its cash cow reputation (mature). As a result, it is difficult to earn more market share because it must be grabbed from rival businesses rather than from new clients. Having a product portfolio like this is incredibly profitable.

(ii) **BCG Matrix Star:** ***High market shares and rapid market growth*** are characteristics of a star type of product portfolio. Stars are the market's dominant product, but they have a lot of work to do to keep their market share advantage. This is due to the market's rapid growth and the opportunity for competitor companies to obtain market share by luring the influx of new clients. Rising stars must therefore invest more in marketing in order to maintain their position. If they succeed, they will reap the rewards since they will become the cash cows once the market has reached maturity.

(iii) **BCG Matrix Question Mark or Problem Child:** This is for product having "*low market shares and high market growth*".

Businesses have trouble with question marks. Although many won't succeed in breaking through and making large profits, there's a chance they'll end up being the future's biggest stars. This is due to the market's rapid growth, which gives businesses the opportunity to increase their market share by acquiring new clients, which is far simpler than luring them away from a competitor. Spending a lot of money on marketing is necessary for a business that wants to expand into a question mark, but even then, success is not guaranteed. Business should invest in the "question marks" with the best chances by carefully selecting which ones to grow.

BCM Matrix Dog: A dog classification denotes a product with ***"low market shares and low market growth."*** Few companies actually sell products that include dogs. In addition to having a large market share, these products have very little prospect of expanding that share because the market is not expanding. Dogs (divest) are typically discarded by businesses unless they have additional benefits, like being a vital part of a successful product line.

2.5 Product Positioning

A sort of marketing called "product positioning" emphasizes a product's benefits to a particular target market. By doing market research and focus groups, marketers can use customer feedback to determine the audience to target.

A business must create and explain a product positioning strategy after determining a segmentation approach. Product positioning is the process of tailoring a company's offerings and marketing strategy to occupy a specific space in customers' minds. When deciding whether to compete similarly to them or try to fill a market gap, the company should conduct marketing research to identify where the products of its rivals stand in each market segment. The last step in the target marketing process overall is product positioning. The company is then in a position to continue planning the specifics of its marketing strategies once it has established its product positioning strategy.

2.5.1 Positioning Method

The steps for product positioning are as follows:

Step 1. *Identifying the differences or positioning concept*: Marketers must comprehend the consumer needs that will drive consumers to buy a product. A marketer can position his product in a variety of ways to increase or improve the value it provides customers. A marketing offer can be differentiated based on the good, services, people, channel, and image. As a result, many strategies for product positioning are created using:

(i) **Product differentiation:** Products can be distinguished based on features such as form, size, color, performance, and many more. For instance, Colgate launched a herbal variant by emphasizing the natural feel that is favored in rural areas.

(ii) **Services differentiation**: Services can be classified according to how they are delivered, installed, and maintained. Examples include lengthy warranty terms, free service coupons, 24-hour services, emergency treatment, etc. In order to distinguish its goods from those of the three well-known public sector companies IOCL (Indian Oil Corporation Limited), HPCL (Hindustan Petroleum Corporation Limited), and BPCL (Bharat Petroleum Corporation Limited), Reliance, a private LPG firm, competed against them. Despite the fact that its items are more expensive, it was able to successfully differentiate them.

(iii) **People differentiation:** Consumers can distinguish the image of goods and services by looking to people or personalities they respect and admire, such as movie and sports celebrities. For instance, Aamir Khan would significantly differentiate the brand and boost sales if he promoted Coca-Cola while sporting a villager's outfit.

(iv) **Image differentiation**: Despite a product being very comparable to one from a rival company, the consumer may choose a brand or company based on its reputation. Advertisements, symbols, signs, colors, and logos are all used to create an image. In the case of rural consumers, extra caution should be used.

Step 2. ***Select the positioning concept:*** The marketer must choose the best and most efficient options based on a variety of positioning criteria. A marketer must choose a positioning idea that acts as a link between the items and the intended audience.

Among the essential factors to consider when positioning a brand are:

(a) **Attractive** - Does it provide the client with value?

(b) **Distinctive** – Is it distinct from those of its rivals?

(c) **Preemptive** – Is it extremely tough for rivals to duplicate it?

(d) **Affordable** – Can customers afford it?

(e) **Communicable** – Can the difference be explained clearly?

Step 3. ***Developing the concept*****:** The marketer must effectively build the concept once the positioning approach has been decided upon in order to properly address the target market. To effectively reach the target market, the marketer must choose the right media vehicle. Marketers should work to make the positioning more appealing to the target audience by bringing it closer to them.

Step 4. ***Communicating the concept*****:** After the concept is developed, the product can be promoted in prestigious journals and periodicals. An effective communication must make the target market, the value proposition, and the supporting product differentiator clear. The positioning for rural markets should emphasize the product's general benefit.

2.5.2 Positioning Strategies

Pharmaceutical companies often use technicalities in positioning a product. Now day's companies are able to understand the relative advantages of a product. Some companies due to their excellent marketing expertise are able to position their me-too products in an effective way.

There are various positioning possibilities to develop the strategies:

Positioning on the ultimate use: Normally in this sector the number of times a patient needs to take the drug a day decides the fate of the drug. It is quite possible that if a drug is to be administered every six hours, there is a possibility that the patient may miss one or two doses. Hence, the alternate routes of administration should be identified for the use of the drug such as once a day or once a week is very easy application possibilities.

Positioning with respect to specific attributes: The specific high points of a product depends on namely efficacy, low side effects and free from contraindications etc., in any product. These mentioned attributes focused more efficiently for positioning the drug in the market. Example Centron from Torrent Pharmaceutical, is positioned as non-steroidal with low side effects and convenient dosage regimen.

Consumer based positioning: Many companies rely heavily on the ease of use, convenience in getting across to the doctors by the consumer types (as infants, children, adult and geriatrics). Example, whenever a geriatric drug is introduced the companies try to have the patient in mind while promoting the product.

Positioning with respect to price: Whenever a price reduction is announced by a company, one can understand that company is playing on the price platform. Often such companies resort to advertising their price in the journal and also in newspapers if it is intended for general practitioners.

Positioning on the packages: Those companies which are operating in both OTC and ethical markets need to differentiate themselves by introducing drugs in novel packages. Glaxo has done the same when they introduced for its asthmatic inhaler in the novel form Disk haler. When the product is highly deleterious for the children, the child resistant packages are being introduced by the companies. Companies could exploit the form packs, pouches, strips which are very effective guard against moisture, alien aromas, light, gases and temperature. Lupin laboratories used packaging effectively in positioning their anti-TB kit AKT-4.

Positioning on diseases: Many companies are trying to be identified as the one in a specific treatment area. Lupin became famous when they started concentrating on the anti-tuberculosis market. They focused their R&D mainly in these markets and have become number one in that area.

2.5.3 Benefits of Product Positioning

1. **Make entire organization market-oriented:** Product positioning is part of the broader marketing idea. It focuses on identifying superior product attributes and more effectively matching them with customers than competitors. This philosophy focuses the company as a whole on the market.
2. **Cope with market changes:** Before the product is successfully positioned, the manager's work is not finished. Market observers must use caution. In order to adapt to changing consumer expectations, new competitive advantages should be identified, created, or acknowledged. The manager becomes dynamic, active, and watchful as a result.
3. **To meet expectation of buyers:** The benefits to be presented are often chosen based on the expectations of the target audience. Therefore, product placement might help to fulfil customer expectations.
4. **Promote consumer goodwill and loyalty:** Systematic product positioning strengthens the company's brand, product, and identity. It improves brand recognition, and the company can foster goodwill and win the loyalty of its clients.
5. **Design promotional strategy:** Greater depth can be added to advertising programming. Based on the benefits that should be highlighted, appropriate approaches are selected to advertise the product.
6. **Win attention and interest of consumers:** The positioning of a product relates to its consumer-important advantages. When these advantages are effectively communicated to consumers, they are unquestionably interested and engaged.
7. **Attract different types of consumers:** Customers' expectations for the products can differ. Some people seek durability, while others seek special traits, novelty, safety, affordability, and so on. A company can attract various clientele by advertising its distinct competitive advantages.
8. **Face competition:** This is the primary use of product positioning. The company has the ability to respond to competitors aggressively. It could increase its competitiveness.
9. **Introduce new product successfully**: When a company releases a new product, product positioning may be advantageous. It can highlight the product's unique and exceptional qualities and enter the market fast.
10. **Communicate new and varied feature added subsequently:** When a company modifies the traits and/or features of its current products, these developments might be contrasted with those of the competitors' wares. Product positioning improves a business's capacity for competition. Typically, consumers consider a product's advantages before buying it. As a result, product positioning demonstrates how an organization's offers are superior to those of competitors. Additionally, it might help clients choose the ideal product.

2.6 New Product Decisions

Product Decision: Price, promotion, and distribution channels—components of the "marketing mix"—are the foundations of any product decision. Decisions about products are arguably the most important because they represent the pinnacle of marketing strategy. It is crucial that the production of the product or service be carefully planned and coordinated, both inside the company and with other functional areas, particularly marketing, when decisions are made about creating or selling goods and services on the global market. The physical product, packaging, labelling, branding, warranty, and service are the primary factors to be considered during the production process.
Key ingredients involved for developing a product, but fundamentally it can be given by series of decisions.

Quality decision = quality product

Problems how we make decisions:

- Limited self-sufficiency
- Imprecision on who makes decisions
- Elusiveness on what to do with feedback
- Ambiguity on accountability

These problems lead to occasionally sluggish decision-making, decisions of inferior quality, and bad experiences for all parties. The basic principles for decision making have been shown in **Figure 2.1** and the pillars of product making decision making have been demonstrated in **Figure 2.2.**

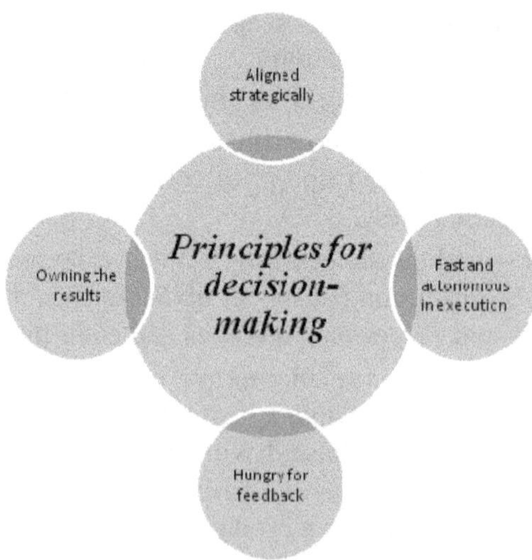

Figure 2.1 Framework of product decision-making

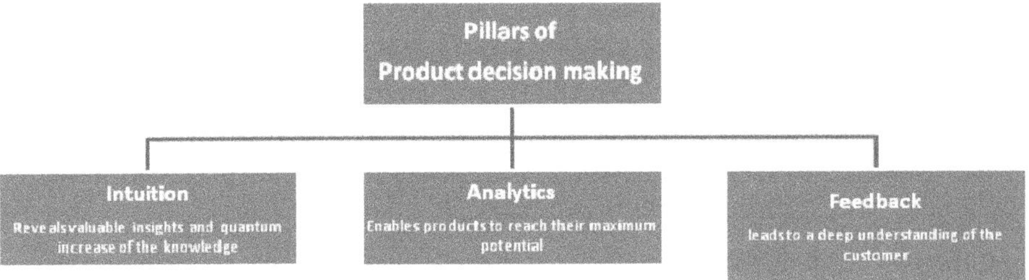

Figure 2.2 Important pillars of product decision making

Elements of new product

Major innovation: A market needs innovations in order to create a new product. They are produced using modern technology and offer clients novel experiences. Examples include the fact that before their first release, phones, smart phones, computers, and tablets did not exist. They attracted potential clients by claiming to make their lives easier if they utilise these products, which led to the creation of new markets as opposed to existing ones. However, there is a danger when trying to draw buyers to a huge innovative product since they might question its value. As a result, the company that developed the new product must find ways to persuade consumers that they need it.

Product improvements: In contrast to "major innovation," products in the "product improvement" group aren't created with the goal of expanding into new markets. Instead, they target customers of competitive companies. The food (diet, fat-free, allergen-free products) and chemistry (particularly detergent products) industries are big users of these novel items. Businesses in these sectors attempt to draw clients to their goods by setting them apart from those of rivals in the marketplace.

Product additions: In contrast to "major innovation," products in the "product improvement" group aren't created with the goal of expanding into new markets. Instead, they target customers of competitive companies. These are fake products that the market purchases from the producers of actual products. The benefits—what customers actually experience in comparison to the original product—will remain limited even though these goods may boast new features. These types of new products are frequently chosen by small businesses because they lack the capacity to create original products.

Repositioned products: Repositioned products are promoted in novel ways to appeal to different types of customers. These are not new products, formulations, or features; rather, they have been given alternative market positionings in order to appeal to diverse client segments. For instance, the Lucozade energy drink firm changed the impression of their product from a drink for people who are recovering from illness to a drink for people who want to exercise.

Technological breakthroughs: These novel products include vaccines for AIDS or cancer as well as cutting-edge technological advancements like flying cars. At the moment of these products' introduction to the market, clients are experiencing something new. Customers receive very diverse or minor benefits from them. The results of ongoing product and marketing research lead to technologically ground-breaking products.

Significant improvements: These product attributes were created through significant market-available product enhancement. This enhancement increases the value of the product, which benefits both consumers and businesses. As an illustration, instant coffee can be used instead of traditionally brewed coffee. While consumers appreciate how much faster and easier it is to make coffee for breakfast, companies improve sales.

Modified products: By adding a new scent to detergent or changing the size of existing products (such as fries or smartphones) insignificantly, this feature allows for the creation of new products.

2.6.1 Stages of New Product Decision

For manufactured goods, the new product choice framework consists of eight crucial elements:

1. ***Idea generation*** is the methodical, continual process of looking for potential new product sources, which may involve improving or revamping an existing product. The goal is to generate ideas for novel products or services or improvements to already existing ones that address market needs.
2. ***Idea screening*** is carried out for less desirable, unprofitable, and undesired products that are currently on the market. Unsuitable concepts should be eliminated after careful study, including early testing and consumer feedback.
3. ***Concept development and testing is vital*** if at this point, customer feedback is used in place of the internal, objective analysis of step two. The idea or product concept now needs to be evaluated by a live consumer base. In light of the testers' feedback, the concept can then be changed and further developed. An illustration of concept development is the idea of automobiles created by automakers. At auto shows, these clay prototypes are on display to get feedback from customers.
4. ***Market strategy/business analysis*** determines the best marketing and sales approach for your good or service. There are four components to it: **product, pricing, promotion, and placement.**
   - **Product**: The product or service that has been created to meet the needs of a target market.
   - **Price**: Pricing decisions have an impact on supply and demand, profit margins, and market strategy.

- **Promotion**: Promoting a product has three main objectives: introducing it to the target market, boosting demand, and showcasing its benefits. Public relations, marketing campaigns, and commercials all fall under promotion.
- **Place**: The location of the product placement must be determined by the targeted potential customers who are located in various regions and where they enter the market to do so.

5. ***Feasibility analysis/study*** delivers information that is crucial to the product's success. It calls for assembling covert teams of testers who will use a beta version—or prototype—of the product and then provide feedback to a test panel. This information reveals the level of interest and desired characteristics within the target market, as well as if the product being developed has the potential to be profitable, marketable, and viable for the company while satisfying a real need among the target market.
6. ***Product technical design/Product development*** combines the findings of the feasibility studies with product user reviews. The objectives of this stage are to turn the concept into a marketable product, alert the departments engaged in the product launch, such as those in R&D, finance, marketing, production, or operations, and coordinate those departments' efforts.
7. ***Test marketing or market testing*** validation of the entire concept, including the marketing message and viewpoint, packaging, advertising, and distribution, is the goal. In test marketing, the product is typically offered to a sample of the target market that is indicative of the entire market. By testing the entire package before launch, the company may evaluate the reception to the product before making a full go-to-market investment.
8. ***Market entry/commercialization*** the product is introduced to the target market at this phase. All the information obtained throughout the first seven stages of this approach is used to design, market, and distribute the finished product to and through the appropriate channels. Now once the product is available to everyone, the "product lifecycle" begins. The response of the target market, the level of competition, and ensuing enhancements to the product offering all have an impact on how long the product will last.

2.7 Product Branding

A brand is described as "a name, word, sign, symbol, particular design, or some combination of these features that is meant to symbolise the goods or services of one seller or a group of sellers," by the American Marketing Association (AMA).

Product branding is a mark or pattern that offers your goods a recognisable brand in the marketplace. These products are distinguished from those of rivals by their brands. A brand is actually a seller's promise to the customer to offer a specific set of benefits or qualities. Each brand embodies a certain level of its attributes, advantages, qualities, beliefs, cultures, and personalities. **Figure 2.3** lists the six features that a brand must guarantee.

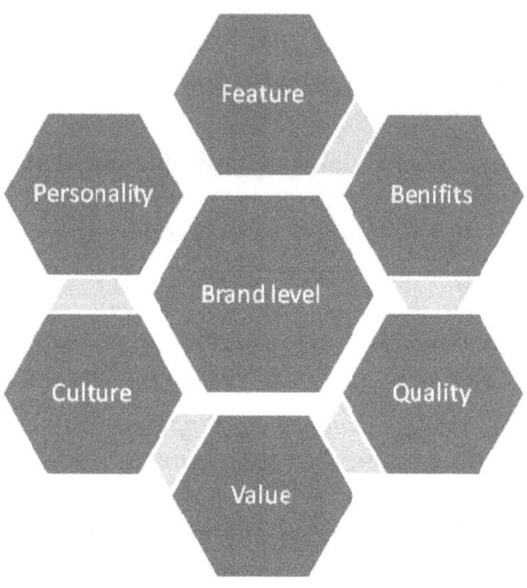

Figure 2.3 Six Features for branding a product

Branding should be done in such a way that others must not be in a position to copy it. Branding of a products means a long term investment in developing a brand by spending heavily on promotion over a period of time. Many companies are known more based on the brands they are marketing. Example; Horlicks, Smithkline Beecham are in the well-known brands of Indian market. A brand name of a drug should be simple and technical for prescription and must elicit a mental image, reflects the products benefits, attributes, positioning and use. Example the use of the drug like Bactrim of NPIL clearly indicates the use as the bactericidal infection fighter.

Brand positioning in pharma sector: Pharma companies must take a few crucial actions in order to overcome the challenges and create a more patient-centric approach to brand branding:

(i) ***Develop one overall brand positioning with the patient at the center*:** A lot of brand teams have distinct HCP (Healthcare Professional) and patient teams that concentrate on their respective target markets. The HCP team may automatically take the lead in developing brand positioning, or each team may design its own positioning tailored to each client type, which results in inefficiency. Instead of creating several positioning statements, create one that is concentrated on the patient's significant benefits, which all stakeholders will be concerned about.

(ii) ***Explore new territories with patients***: In terms of brand positioning, the pharmaceutical industry often emphasized therapeutic benefits (efficacy, safety, dose, etc.). Few people truly take the time to comprehend the main advantages for the patient and how it will affect their lives. Conduct patient research before and during the development of your brand positioning to understand the impact of their disease and your product in their own words (as opposed to the physician's assessment of the impact). This will guarantee that you are incorporating the voice of the patient. As a result, it will be possible to create a successful brand positioning using significantly more facts.

(iii) ***Link the positioning to customer-specific benefits and messaging***: While the patient should be the center of attention for the brand positioning, other stakeholders must also benefit from the outcome. As a result, a value proposition and message tailored to individual customers will be created and tied to thc positioning's primary differentiating benefit. For instance, if a medication is simple for patients to use, one may emphasize how this can help HCPs be less concerned about patient compliance or correct technique. In reality, these physician- or practice-oriented benefits, together with factors like the reputation of the pharmaceutical firm, can make a considerable difference in your brand's favor in situations where the product features and functional benefits of two medications are comparable. Another example is the arthritis medication Voveran from Novartis, which is marketed as having traditional dosage guidelines and gastro intestinal side effects of reduced intensity.

2.7.1 Characteristics of a Good Brand Name

A good brand must have following features and while selecting a brand name few opinions must be considered:

1. **Short and simple:** The chosen brand name should be brief, straightforward, and simple to pronounce, remember, and spell.
2. **Relevancy**: A strong brand must be current, live up to consumer expectations, and function as intended. Although the product is distinctive, it may not sell if a poor job isn't done to convince them to buy it.
3. **Consistency**: Consumer trust in a company is increased by a consistent brand that conveys what it stands for. When a company expresses its message in a way that stays true to its primary brand proposition, it has a consistent brand.
4. **Proper positioning:** A powerful brand should be established in such a way that its target market remembers it and favors it over rival goods.
5. **Sustainable**: A successful brand is beneficial to a business. A sustainable brand propels a company toward success and innovation. One example is the sustainable company Marks and Spencer's.

6. **Credibility:** A strong brand should fulfil its commitments. You should be realistic in how you present your brand to the public and customers. It shouldn't fall short of keeping its commitments. Customers want to believe in the promises you make to them, so avoid being overly dramatic.
7. **Inspirational:** A brand should expand beyond its niche and serve as an inspiration. Nike Phenomenal Jersey Polo Shirt, as an illustration.
8. **Uniqueness:** A brand ought to be distinctive and distinguished. You should be distinguished from other market competitors by it.
9. **Appealing:** A brand should be appealing, and the promise you make and the value you provide should draw the buyer in.
10. **Capable**: The brand name should be capable of being registered and of staying in the market for a longer period of time.
11. **Adaptable**: The names selected should be adaptable to packaging or labelling requirements, different languages etc.
12. **Versatile**: The brand name is required to be versatile so that a new product can be easily accommodated.

Advantages of product branding:

1. **Public Identification**: Your advertising and marketing activities can greatly benefit from effective product branding. Utilizing the same identity package (logo, design, packaging, etc.) consistently and communicating the same message can help to elevate the quality of your brand so that it is easier for consumers to recall and recognise. A certain level of public recognition may enable investment in other company sectors by lowering marketing expenses.
2. **Enables marking product differentiation**: Branding enables a marketer to distinguish its products from its competitors' products. Hence; it ensures secured and controlled market for a product.
3. **Preference:** Strong brands advance in consumer preference over rivals, giving them a competitive edge. When presented with a variety of brand options for a certain product, buyers are more likely to select well-known and reputable brands. Or to put it another way, having a strong brand helps to retain customers.
4. **Helps in advertising and display programs:** Branding helps in effective advertising and display programs.
5. **Market extension:** When the reputation of your product is well-established, you might consider expanding its potential for diversification or breaking into a new market. Because they have already provided their approval of the product and brand, customers who are already familiar with them will be more willing to support the market expansion.

6. **Barrier:** A strong product brand can also break down barriers to entry for new competitors and aid in market share projection. People that wish to compete in your industry must invest a lot of money in building their brands to meet your criteria. Additionally, a new pricing threshold can be established for clients who are adamant about sticking with your brand because they are more likely to choose your products than other, less expensive options.
7. **Differential pricing**: Branding helps in charging different prices for products because when the customer becomes habitual or used to any particular brand of a product, they don't mind paying higher price for it.
8. **Ease in introduction of a new product:** When a new product enters the market with an established brand name, it is immediately recognizable. Many companies have adopted this strategy of introducing a new product under their established brand name.
9. **Ensure quality**: Branded product assure good quality to the customers.
10. **Status symbol:** The customers who use branded products feel proud and satisfied.

2.7.2 Brand Name Selection

The brand name must convey an image of the product. To develop a successful brand name, Philip Kotler offers a number of recommendations:

- The brand should allude to a feature or benefit of the product, such as a colour or an activity (e.g., Frooti, Appy, Fizz).
- It must be simple to say, recognise, and remember, like Amul, Tide, Kissan, and Lux soap, for example.
- Have distinctive nature as cello, VIP
- Must be expandable.
- Like Mr. White, the name ought to be simply translated into other languages.
- Must not carry poor meaning in multiple language.
- Be able to register and receive legal protection.

Choosing a brand name process:

- **Define the objectives of branding in terms of six criterions**: as being detailed, evocative, complex, traditional, arbitrary, and fantastical. Understanding a brand's function within a corporate branding strategy as well as its relationship to other brands and products is crucial. Additionally important is comprehending how the brand fits into the broader marketing plan, and a thorough analysis of the specialised market must be taken into account.
- **Generation of multiple names**: You may obtain names from any source, including the company, its management and staff, its existing or prospective clients, its agencies, and its professional consultants.

- **Names are screened based on branding goals and marketing considerations:** The brand names must, among other things, not have any bad connotations, be easy to pronounce, and abide by all applicable laws.
- **Compiling extra information about each of the chosen names:** There should be an exhaustive global legal search. These searches are periodically carried out in order to reduce costs.
- **Conducting consumer research:** Consumer research is commonly conducted to support management expectations for brand recall and importance. The features of the product, its price, and any promotions might be displayed to customers to help them comprehend the value of the brand name and how it will be used.

Management can pick a brand name that will best serve the organization's branding and marketing objectives based on the aforementioned stages, and formally register that name.

2.7.3 Brand Extensions

Brand extension involves transferring a brand name from one product category to another. When the new categories are conceptually connected to the old in the minds of the consumers, brand expansions typically perform well. Using a brand extension has a variety of implications:

(a) The cost of introduction is much lower because consumer education is not needed.

(b) Advertising and other promotional expenditure are more effective because they produce greater awareness and recognition of the brand.

(c) The risk of introducing the product is reduced because the new products image is already established

(d) The development costs are reduced and the cost of products failure is lower.

Examples: The NASIDS, Diclofenac sodium is branded Voveran by Novartis. The company extended the brand into the rubefacients and tropical analgesic segment as Voveran Emulgel with good results. Another example of a broad spectrum Povodine-Iodine was branded Betadine by Win-Medicare and it extended the brand into the surgical as well as mouthwash segments

2.7.4 Brand Building Process

Brands once introduced into the market must enable the buyers to identify it, remember it and if satisfied with its performance purchases it. Most potential buyers go through various steps before they become brand loyal. The brand loyal comes with:

(a) Brand Recognition and awareness: When a brand is recognized the consumer has seen it before either through the prescription of the doctor or at point of purchase. A recognized brand is one that has successfully implemented its image in the consumer's mind.

(b) **Brand acceptance and Rejection**: Brand acceptance means that a brands image is received favorably by the doctors and the ultimate consumers or outcome of the preference of the brand name in the mind set of consumers. Brand rejection means that consumers are either unreceptive to the image or unsatisfied with the product. If brand rejection is the result of an image problem marketer will have to either change the brands image or target a more receptive segment of the population. If the problem is the rejection of the product itself then either the product or the methods used to produce will have to be changed.

(c) **Brand preference**: Consumers have arrived at the stage of brand choice when they seek a particular brand. Buyers who are convinced of a brand's value would prefer to use it above any of its rivals are said to prefer it. Therefore, differentiating a brand from a competitor's product is a fundamental strategy for fostering brand loyalty.

(d) **Brand loyalty**: Branding adds value but only over time. It takes years to develop a recognized brand and make it valued by a group of consumers. It ensures that consumer will develop a relationship with the company's brand and buy that brand repeatedly or to retain the existing customer on a continuous basis. It is less costly to do so than to attract new customers. Even when the new customer is being lured, the existing customer should not be ignored. Strong brand loyalty is an effective entry barrier for the competitors. Example, TB drugs offered by Lupin laboratories and the doctors tend to be more loyal to the brands from the company. This also helps the company in pushing the other common drugs easily.

(e) **Brand Association**: Consumers associate the brand with certain attributes, a celebrity endorser or a visual symbol. Example: The boy used in handy plast advertisement, the fingers shown for the Vicks Action 500+ to create association, Anacin which had excellent brand association with four spread out fingers.

2.8 Packaging and Labelling Decisions

Packaging: The science, art, and technology of packaging involves confining or safeguarding goods for distribution, storage, sale, and use. Because a product's package is a selling tool, developing it is a crucial component of product planning and promotion. In particular for non-durable consumer goods, it is essential to the success or failure of many items' marketing initiatives. It protects the product while also acting as a marketing tool.

The process of creating and designing a product's package or container. A product's packaging is crucial to its marketing. Some buyers in remote areas recognize a product by its design or packaging and decide to purchase it. Packaging has the potential to be an effective tool for building brand or product loyalty. Packaging has a crucial strategic function to play in fostering consumer acceptance of a brand or product since it is a consistent and permanent statement of the brand. According to ***Charles W. Lamb***, packaging has four distinct marketing functions:

- Protects the product.
- Promotes the product.
- Use the product
- Packaging the product for reducing the environmental damage.

The main objectives of packaging are as follows:

- **Product storage:** The package serves as a container for the product. Examples include shampoo sachets and bottles, cooking oil bottles or tins, milk packets, and so on.
- **Product protection:** The packaging shields the product from heat, sunlight, and moisture. Additionally, it shields the item from leaking, deterioration, and breaking.
- **Providing information to the customer:** Product packaging gives customers detailed information about the product's name, manufacturer, contents, price, date of manufacture, and expiration date, as well as usage instructions.
- **Product differentiation:** Product differentiation is created through packaging. Customers can quickly distinguish one product from another by looking at the packaging's design and colour.

Requirements of good product packaging are:

- It should be safe to handle and transport.
- Bulk packs may come with handling features like hooks, handles, grippers, etc. to make handling easier. It should be easy to handle.
- Must adapt to a fast assessment of the information. Keep in mind that both the exporting and importing countries' customs authorities may want to inspect the contents.
- The manufacturer must be able to quickly recognise it from the package.
- The specific contents must not be shared unless absolutely essential.
- It ought to be simple to get rid of.
- The packaging must adhere to the buyer's requirements. Additionally, it must adhere to the laws of the exporting and importing nations, the rules and regulations of the shipping company, etc.

Features of Packaging are:

The packaging of product must have the following features as:

- Should be eye-catching, colourful, and visually appealing packages that can convey significant messages about the functionality, features, and advantages of the product.

- Sometimes a box will contain several complementing goods. The process of merging numerous goods and services into one bundle is known as mixed bundling. Price bundling occurs when two or more products are offered for sale at a single bundle price.
- Packages frequently work to project an image of importance, distinction, or convenience. Additionally, they provide usage instructions and details on the product's ingredients, nutritional value, product promises, and potential risks.
- To meet the needs of various market sectors, including both individuals and families, packaging is available in a variety of sizes.
- Packaging can contribute to the customer's increased product safety. Many goods that were previously wrapped in glass are now available in plastic packaging. To ensure the safety of the consumer, goods such as non-prescription medicines, cosmetics, and food are sold in tamper-proof blister packets.
- The product must be packaged correctly and securely to protect it from tampering, contamination, leakage, evaporation, spoilage, or damage while it is being kept, transported, or advertised.
- It has led to a rise in the use of packaged goods, which lowers the likelihood of adulteration.
- Given the information from self-service shops, packaging currently has a position of silent salesmanship, particularly at self-service stores.

Levels of packaging (Figure 2.5):

1. **Primary package**: The material that at first encases and holds the goods. This is typically the package that interacts directly with the contents and acts as the smallest distribution or use unit. Ampoules, Vials, Blister packs, Bottles, and Sachet packaging are a few of the primary pharmaceutical packaging styles that are frequently used, as shown in **Figure 2.4**.

Figure 2.4 Primary pharmaceutical packaging

2. **Secondary packaging**: Outside of the primary packaging, secondary packaging may be used to group primary packages together. Examples include injection trays, shipping containers, cartons, and boxes.
3. **Tertiary packaging:** Packages necessary for storing, identifying, or transporting are referred to. Examples include barrels, containers, edge guards, etc.

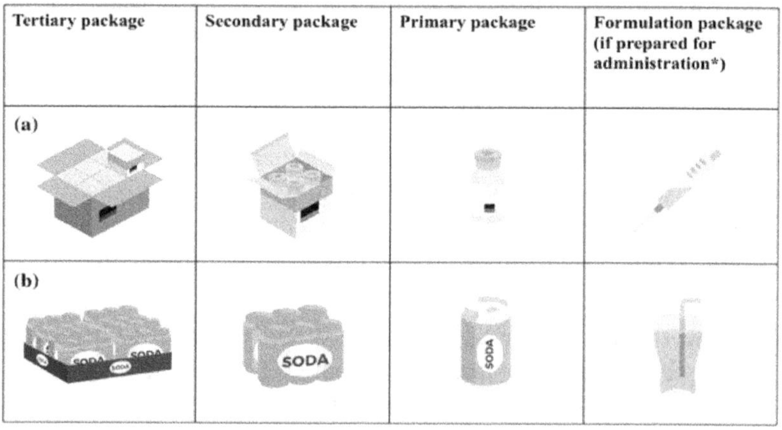

Tertiary package	Secondary package	Primary package	Formulation package (if prepared for administration*)
(a)			
(b)	SODA	SODA	

Figure 2.5 Levels of packaging

After packaging of any product the next is labelling of a product. Since, both have few basic differences which are given in Table 2.2.

Table 2.2 Difference between Packaging and Labelling

Packaging	Labelling
Meaning	
An art of creating a suitable package to wrap or enclose the product in order to protect it from harm and contamination and to make its storage, transportation, and use easier.	providing information to consumers to aid in their purchase decisions Additionally, it is done to comply with legal obligations.
Objective	
Store, safeguard, maintain the product, and establish brand identity.	Communicate with customers.
Focus	
Product presentation	Product description
Design	
Innovative and attractive	Simple and formal

Labeling: For identification and to provide comprehensive information about the goods, printed information is bonded to the item. It contains details on the seller and the goods. The process of labelling involves applying labels to products.

Functions of labeling: The main functions of labelling are as follows:

1. **Complete description of the product:** It describes the item and what's inside: A product's label provides information on how to use it and any precautions that should be taken. Example: The label of Red Label Natural Care tea lists five components that boost immunity.
2. **Identification of products:** Aids in picking out the product from among the many others on the market. The brand name of a chocolate, for instance, will make it easier for someone to select it from the other available confections.
3. **Grading of products:** Labelling helps in grading the product into different categories.
4. **Assists promotion of products:** Explains the benefits of the product to the customer. In the packet of potato chips, for instance, it grabs the consumer's attention by displaying slogans like "20% free" or "save Rs. 15.".
5. **Promotion of products:** Labels that are meticulously created draw in customers and encourage them to buy the product.
6. **In compliance with the law:** Labels must exactly follow the law. For instance, the warning "Tobacco is dangerous to health" should be on the label of cigarettes. On the box of cigarettes are necessary disclaimers like, "Smoking is dangerous to your health.".

Types of label

Brand level: The package only mentions the brand name. Examples include the Dharwad mangoes packaging and the Lux, Hamam, Rexona, etc. soap labels. The only name in bold is the brand.

Grade level: Certain products have been graded. This kind of label indicates the product's grade. It displays the product's quality using words, letters, or figures. Peas that are canned can have an A, B, C, or D grade. Similar to that, the grade label for packed wheat used in fertilizers might list 1, 2, 3, and 4 grades. 19-19-0-19, 17-19-19-19. Some businesses may name their items as good, better, best, etc.

Descriptive level: provides comprehensive information on the product, use, and maintenance. An illustration would be the booklet found within the Vasemol hair dye packet. What safety measures should one take, etc.?

2.9 Product Management in Pharmaceutical Industry

Product management is an organizational position within an organization that deals with new product invention, business justification, planning, verification, forecasting, pricing, product launch, and marketing of a product or items at all stages of the product lifecycle.

In the pharmaceutical sector, managing the organization's current product line, building brands, launching new products, and developing pipeline product strategies are all primary responsibilities of product management. It combines marketing, sales, product development, and business.

In order to achieve revenue and customer satisfaction throughout the Product Lifecycle, the Product Manager is responsible for compiling and prioritising product and customer needs, developing the product vision, and working closely with engineering, sales, marketing, and support. In order to provide patients with life-saving medications and therapies, a product manager is crucial. Pharmaceutical, medical device, and biotechnology companies, like any other sector, have sales and marketing responsibilities for informing the medical community about their offerings.

Because it organizes and executes a company's product marketing strategy, product management plays a crucial role in the pharmaceutical sector. Pre-launch, launch, and post-launch marketing activities are all carefully handled by the vertical headed "Product management in the pharmaceutical sector."

The four stages of a product's life cycle—introduction, growth, saturation, and decline or steady phase—as described in section 2.3—are experienced by all products in the pharmaceutical industry. In all 4 stages of the product life cycle, product management in the pharmaceutical sector is crucial.

The product management hierarchy in the pharmaceutical industry, depending on the size and income of the business, the hierarchical structure of product management in the pharmaceutical industry may change. The basic hierarchy employed in the pharmaceutical industry is shown in Figure 2.6.

Figure 2.6 Product management hierarchy for sales and marketing in the pharmaceutical industry

Product management responsibilities and roles:

In pharmaceutical sector, managing the organization's current product line, building brands, introducing new medications, and developing pipeline product strategies are all important responsibilities of product management. The positions connected to product management in the pharmaceutical sector are listed below:

- Educating the field sales team about the product and giving them guidance so they have both scientific and communication skills.
- Collaborate with cross-functional teams including the organization's R & D (Research & Development) team, learning and development team, and medical affairs team.
- Creating brand plans and strategies for the product line, as well as market penetration plans, competitor analyses, and market research.
- An analysis of the product line's strengths, weaknesses, opportunities, and threats (SWOT) that enables the sales team to seize opportunities and boost product sales.
- Producing brand-related promotional materials such as visual aids (VA), leave-behind literature (LBL), newsletters, flipcharts, or digital campaigns such as the launch of a website or an app or a webinar series, etc.
- Holding conferences, scientific symposia, CMEs (Continuing medical education), meetings, and other events to promote brand awareness among hospitals and healthcare professionals, who make up the pharmaceutical industry's target market.
- Creating training camps, award ceremonies, and recognition programmes to encourage the sales team.
- Product forecasting, new product pipeline tactics, new product pre-launch, launch, and post-launch tactics, as well as fresh initiatives for product growth tactics.

Skill sets of Product Management in the Pharmaceutical Industry

- Product expertise, as well as creativity and analytical abilities.
- Market research and competitive analysis intelligence.
- Excellent communication abilities.
- The capacity to collaborate and work with cross-functional teams.
- Building a team.
- Forecasting and a creative strategy.
- Leadership and management abilities in sales.

Significance of Product Management:

- The sales team and product management are in charge of helping a business reach its top- and bottom-line revenue goals.
- In charge of creating, consolidating, and establishing brands.
- Establishes a framework for the market penetration plan and new product introduction, and sees to its execution.
- The Product Management team establishes a plan for product establishment in the pharmaceutical business by building strong relationships with and coordinating with medical staff and hospitals, and then ensures that the sales team puts that strategy into action.

Answer the Following Questions

Q1. Define the term product and write its classification.

Q2. How can we group products according to their use, tangibility, and durability?

Q3. Write a note on pharmaceutical products.

Q4. Define product line and multiple product line.

Q5. Discuss product line extension with suitable example.

Q6. Discuss product line stretching with suitable example.

Q7. What distinguishes product line pruning from product line filling?

Q8. Discuss product mix with suitable example.

Q9. Enumerate the dimensions of company's product mix.

Q10. Discuss the various phases of product life cycle.

Q11. Define product portfolio and write its importance.

Q12. How you will carry out the product portfolio planning?

Q13. Define product positioning and enumerate the steps involved in it.

Q14. Enlist the benefits of product positioning.

Q15. What are the steps taken by pharma companies towards patient-centric approach for brand positioning?

Q16. What are the elements of product decision?

Q17. Write the various stages of new product decision and discuss any three.

Q18. What is Product branding, mention its advantages?

Q19. Write about brand positioning in Pharmaceutical Industry?

Q20. Differentiate between Packaging and Labeling.

MCQs

1. Who suggested a company's "Market Mix" should include its product, pricing, location, and promotion?
   (A) Philip Kotler (B) Stephen Morse
   (C) Nielsen (D) Neil Borden
2. The 7Ps in the marketing mix do not include this P?
   (A) Purpose (B) Promotion
   (C) People (D) Price
3. It is the most popular strategy for any business to generate leads.
   (A) White pages (B) Blue pages
   (C) Yellow pages (D) Green pages
4. It's called "using well-known brand names to promote a new or modified product in a different category."
   (A) Line extension (B) Brand extension
   (C) Multi branding (D) Co-branding
5. What distinguishes a brand is how the consumer thinks about it.
   (A) Brand attitude (B) Personality
   (C) Brand loyalty (D) Brand name
6. What aids in recognizing the item or brand and describing various aspects of the item.
   (A) Store branding (B) Supplying
   (C) Packaging (D) Labeling
7. Used to distinguish a seller's goods or services from those of the competitors, names, signs, symbols, designs, etc.
   (A) Branding (B) Co-branding
   (C) Mega branding (D) Store branding
8. Which concept entails introducing fresh or altered goods in a new class or line while leveraging a powerful brand name?
   (A) Brand dilution (B) Brand extension
   (C) Brand bonding (D) Brand elements
9. The term "intangible" refers to goods and services that cannot be seen, tasted, touched, heard, or smelled before being purchased.
   (A) Brand testing (B) Service intangibility
   (C) Support service (D) Brand strategies

10. Which one helps in identifying the product.
    (A) Brand (B) Label
    (C) Trademark (D) Packaging
11. Same brand name but different product is called
    (A) Line extension (B) Tagline
    (C) Line stretching (D) Brand promotion
12. Manufacture sells his products his product using its own brand called
    (A) License brand (B) Co-branding
    (C) Manufacturer's brand (D) Private brand
13. What does the acronym PLC mean?
    (A) Product life cycle (B) Production life cycle
    (C) Product long cycle (D) Production long cycle
14. The product's size, color, brand, and packaging are all taken into consideration.
    (A) Chemical features of product (B) Physical features of product
    (C) Product designing (D) Product manufacture
15. A stage in the "Product Life Cycle" represents the quickly increasing product sales and is known as
    (A) Market introduction phase (B) Growth phase
    (C) Saturation phase (D) Mature phase
16. Povodine -Iodine was branded Betadine by
    (A) Win-Medicare (B) Cipla
    (C) Sun Pharma (D) Mankind
17. Vasemol is
    (A) Powder (B) Face pack
    (C) Lip color (D) hair dye
18. LBL extends as
    (A) Leave behind literature (B) Left behind literature
    (C) Leave behind list (D) Leave behind leaflet
19. Zintetac was launched in the international market by
    (A) Win-Medicare (B) Cipla
    (C) Glaxo (D) Mankind

20. Diclofenac sodium is branded Voveran by
    (A) Win-Medicare (B) Cipla
    (C) Glaxo (D) Novartis
21. Anticoagulant 'Warfarin' was sold under the brand name
    (A) Providin (B) Provigil
    (C) Coumatin (D) Coumadin
22. In FDC "C" is the representation of
    (A) Cosmetic (B) Care
    (C) Committee (D) Coumadin
23. The drug registration procedure was first started in which country by FDA
    (A) USA (B) China
    (C) India (D) Japan
24. Idea screening is based on
    (A) Attractiveness index equation (B) Positive index equation
    (C) Effective index equation (D) Valuable index equation
25. Lupin laboratories used packaging effectively in positioning which of the following
    (A) Anti-TB kit (B) Anti-Viral kit
    (C) Anti-Malarial kit (D) Anti-Cancer kit
26. Fast moving consumer goods (FMCG) comes under the
    (A) Convenience products (B) Shopping products
    (C) Speciality products (D) Unsought products
27. Which has a high market share and high market growth.
    (A) BCG matrix Star (B) BCG matrix Dog
    (C) BCG matrix Cash flow (D) None
28. BCG in product decision stands as
    (A) Boston Career Group (B) Boston Consulting Group
    (C) Basic Consulting Group (D) British Consulting Group
29. From the below mentioned option which is not a dimension of product portfolio
    (A) Amplitude (B) Consistency
    (C) Depth (D) Width

30. INH is a drug of choice for
    (A) Tuberculosis (B) Cancer
    (C) Diabetes (D) Malaria

Answer key to MCQs

1. D 2. A, 3. C, 4. B, 5. A, 6. D, 7. A, 8. B, 9. B, 10. A,
11. A, 12. C, 13. A, 14. A, 15. B, 16. A 17. D, 18. A, 19. C, 20. D,
21. D, 22. A, 23. A, 24. A, 25. A, 26.A, 27. A, 28. B, 29. D, 30. A

CHAPTER 3

Promotion

3.1 Method for Promotion

The common methods of promotions are:

1. **Advertising:** An aggressively commercial, impersonal message is used in advertising to sell or promote a product, service, or idea. It is a paid commercial meant to influence consumer choices. Ads do this by using emotive language, which is meant to elicit a certain emotion, such as joy, sorrow, or terror. One example of a phrase that creates urgency is "buy it before it's too late.".

   Objectives for Advertising:

   - Creating information for consumers of a new product or service.
   - Creating an environment that encourages customers to purchase a good or service;
   - Reminding customers of the advantages they may expect from a good or service.

   There are various advertising channels for promotions such as television channels, radio, leaflets and flyers, reprints of ads, display signs, newspapers, websites, blogs etc.

   The media outlets utilised to sell a business or a product are extremely important. Investing in national paper advertising, for instance, would cost a tiny local firm money even if they would profit from it. A company may elect to utilise social media rather than traditional forms of advertising to target particular clients.

   Advertising in pharmaceutical product marketing includes preparation of product literature, product labels, leaflets, visual aids, advertising in pharma and medical journals, souvenirs, seminar, medical symposia and various print media (local, national), in case of OTC products by radio and TV channels.

2. **Sales Promotion:** Sales promotion is a part of the company's broader promotional strategy in which a number of short-term customer-focused strategies are employed to elevate the product's perceived value and/or attractiveness and so boost demand. Or Sales promotion is a form of marketing strategy in which the product is promoted using rapid, enticing

efforts to increase demand and raise sales. This strategy is frequently used to introduce new products, sell out existing stock, attract more customers, and temporarily increase income.

One kind of sales marketing is special offers. They provide momentary incentives to entice customers to buy.

Objectives of sale promotion

- To increase the sale volume of product and generate new customers.
- Fasten the sale of slow-moving products and have a clearance of excess products in stock.
- Launch of new products, which will encourage the dealers to actively participate in sale promotion by techniques like the scarcity principle, offers, and penetration pricing.
- Businesses fight with competitors' short-term marketing strategies by utilising temporary sales promotion approaches and highlighting unique product qualities.

In any market there are various ways of sale promotion as discounts, competitions, free gifts, loyalty cards, coupons, free trials etc.

Importance of Sales Promotion: Sales promotion is a successful tactic for hitting short-term sales goals since it entices potential customers to buy the products. It includes important marketing strategies for:

- Expanding the brand's reach to new markets or customers.
- Preserving sales volume and meeting immediate sales goals.
- Stimulating demand in the short run by making the item seem like a fantastic bargain

Sales Promotion Methods:

A. Sales promotion tools used for consumer-oriented promotion are:

Free samples: Giving away free samples promotes a brand and activates the psychology of ownership, which leads people to select the advertised product if they like it.

Free presents: Giving out free items draws clients since they gain more for their money.

Discounts/coupons for discounts: A coupon is a document that entitles the owner to a certain discount when buying a specific goods. The coupons may be given out by a company via the mail, newspapers, magazines, stores, etc. Holders of vouchers can purchase the goods from the stores listed by the firm, for instance Vishal Food Mart, at a discounted price. Discount coupons are a great technique to temporarily increase sales. People utilise coupons for discounts so they can buy things they would otherwise not be able to.

Exchange programmes - Customers are drawn to exchange programmes because they receive some value even for their used goods. When old items are exchanged, the price of a new product is decreased.

Take your outdated colour TV with you, and for Rs. 15,000, take home a Samsung LCD TV.

Finance options: Financial options, such as no-cost EMIs and low-interest EMIs, make it simpler for buyers to buy expensive goods.

Shipping plans: Exorbitant shipping charges might occasionally deter buyers from purchasing things. These expedited delivery plans save friction.

Discounts on packages: These deals are a great way to get rid of surplus items. It implies charging less for items in packages than when the same variety is bought separately.

Bulk purchase deals: This is a fantastic sales marketing strategy to get rid of unwanted stock. It entails giving discounts to clients who make large purchases.

Specific programmes: Various health related programmes are done by the firms to induce sales along with government like Immunization programmes, Pulse Polio programs and National diabetic week.

Multiple promotions: Multiple promotional offers are included in the programme. Example: When buying a one-liter pack of Vatika herbal shampoo, you can get Rs. 30 discount and a multifunctional jar for free. Promotional offer of a free or additional (same or different) product on the product is known as an add-on campaign. Using pears weighing 75g, a complimentary 8ml Sunsilk shampoo sachet is offered.

Combination promotions: A discount is offered on the purchase of two or more goods when they are purchased in a combination pack. Save Rs. 10 when buying Colgate toothpaste and a toothbrush at the same time.

Volume promotion: Offers extra product quantities for free when you buy one. Get 50 more grammes when you buy 100g of Tide, for instance.

Fair and exhibitions: Pharmaceutical companies present their goods and spread awareness of them at various medical and pharmaceutical conferences. Businesses provide complimentary material as an introduction to themselves and their products.

Price promotion: The company provides a product price decrease. For instance, Brooke Bond offered Rs. 5 off when purchasing Revive 200g and Rs. 10 off when purchasing 500g to entice customers to purchase half-kilo packs.

Contest promotion: In order to enter the contest and win rewards, the company asks that the client buy the product. Using Britannia as an example, ***"Britannia khao world cup jao"***.

B. Sales promotion techniques targeted to traders:

Point of purchase displays: To help businesses raise their sales, this involves giving out free point-of-purchase (POP) display devices to them.

Trade shows: Trade exhibitions are a fantastic way for businesses to showcase their products to the thousands of attendees who are also dealers. Additionally, trade exhibitions provide even larger reductions than regular retail prices.

Push money: Extra payments are made to traders as part of this strategy to encourage them to reach predetermined targets. Giving them, for instance, a $30 bonus for selling product B and a $50 bonus for selling product A for a set amount of time.

Deal loaders: These are the presents offered to retailers and wholesalers in return for placing a certain order amount.

Trade deals: These are unique incentives given to retailers to entice them to advertise a certain product and boost its sales for a finite amount of time.

Buying allowances: When buyers request a certain item, vendors receive special discounts. **3.**

3. **Direct marketing:** Without using any middlemen, direct marketing involves advertising and selling goods or services directly to the customer. Direct marketing may frequently increase client numbers; nevertheless, it frequently requires extensive promotion before a sale is made. Numerous direct marketing strategies are frequently employed by businesses, including:
   - **Direct mail**: Send direct mail that visually stands out. The more eye-catching, the more likely busy HCPs (Health care professionals) and pharmacists are to see your mailer—and open it.
   - **Telemarketing**: Direct selling of products or services to prospective customers over the phone or online is what it is.
   - **Mobile marketing of products**: Mobile marketing is a multi-channel, digital marketing strategy that connects with a target audience through their smartphones, tablets, and/or other mobile devices utilising websites, emails, SMS and MMS, social media, and applications.
   - **Personal selling**: It is face-to -face interaction with one or more suitable purchasers for purchase orders. In pharmaceutical industry- medical representatives details the product of the company to doctors with the help of literature and visual aids
   - **Product catalogues and Home shopping marketing are also the part of direct marketing.**
4. **Branding:** Businesses usually work to establish a positive brand image since it may be used for advertising. When a new product is released under an established brand name, consumers may be more likely to purchase it since they are already familiar with the established brand.

   Building a promotional strategy: To reach the target demographic, a promotional plan may use some or all of the aforementioned promotion strategies. The marketing plan will be determined by the size of the company, the amount of money available for advertising, and the target market groups. A successful marketing plan for a small neighbourhood gardening shop can include printed leaflet advertising along with a welcome discount for new clients. This tactic would target local residents, as they are the most likely to become clients. The steps involved for good promotional strategy has been given in **Figure 3.1.**

Figure 3.1 Steps for good promotional strategy

3.2 Determinants of Promotional Mix

The promotional mix is a combination of several promotional strategies used by businesses to keep their products in customers' thoughts and increase demand for them. These strategies include advertising, personal selling, direct marketing, media affairs, and direct marketing. When choosing the promotion mix, management must take the following into account:

Main factors influencing promotional mix has been briefly discussed as under:

1. **Type of product**: When deciding on the promotion mix, it is essential. Products may be categorised into groups like essentials, luxury items, new things, non-branded products, necessities, and so on. All of these products need different types of marketing strategies. Advertising is acceptable, for instance, for very well and branded stuff. For products without branding, selling could be suitable. For a freshly introduced product to get swift consumer approval, four tactics are used: publicity, direct marketing, personal selling, and advertising.

2. **Use of product:** Products may be utilitarian, consumable, essential, or luxury, which affects the media and advertising strategies used. Examples include the widespread use of sales promotion and advertising strategies for consumer goods and the use of personal selling strategies for industrial items.
3. **Complexity of product:** The intricacy of the product has an influence on the promotional strategy selection. Personal selling is more effective for difficult, intricate, risky, and recently created commodities that need in-person explanation and inspection. On the other hand, advertising works best for simple, manageable products.
4. **Purchase quantity and frequency:** The business should consider these factors while deciding on the marketing mix. For things that people buy frequently, advertising is usually used, but sales promotion and personal selling are preferred for goods that people buy more frequently. Heavy and light users are served via advertising and personal selling, respectively.
5. **Funds available for marketing promotions:** The financial health of the organisation has a big influence on the promotional strategy. When compared to sales promotion and personal selling, which are far less expensive strategies, businesses with limited finances cannot afford to advertise on broadcast, television, in papers, or in magazines. In spite of this, the company may choose to exploit some commercially major moments as PR opportunities.
6. **Market type:** The type of promotional mix relies on the customer or market characteristics. Education, location, wealth, personality traits, knowledge, negotiation skills, occupation, age, sex, and other factors are significant determinants of a company's promotion strategy.
7. **Market size:** It stands to reason that a small market makes personal marketing more effective. Markets with a large number of buyers are more suited for advertising. Location is also very important. Geographical regions have an impact on the type of message, language, and sales promotion methods, among other things.
8. **Stage of product life cycle:** A product's life cycle includes four stages. Threats and opportunities fluctuate depending on the stage. Separate marketing tactics are required for each level. Every promotional tool varies in how well it matches different product life cycle stages.

   We can draw the following conclusions:

   1. Sales promotion, personal selling, and advertising are employed at the introduction stage. However, advertising is given more prominence, and during the second stage, more extensive advertising and sales promotion strategies are deployed.

      The corporation decides to limit expenditures while lowering promotional efforts in the fourth stage, after which more aggressive marketing and personal sales are employed in the third stage.

9. **Level of competition:** Promotional campaigns are developed based on the nature and intensity of the competition. Every marketing strategy aims to protect a business' interests against competition. The amount of competition affects the extent of promotional activities and the choice of tools.

10. **Promotional objectives:** It is the primary factor that affects the promotional mix. Various promotional mix tools may be utilised to achieve a variety of objectives. Advertising is wise if a business wants to educate a lot of clients. A business may employ personal selling to win over a select group of clients. The sales promotion may be used even when a business wants to sway clients for a certain occasion or season. Some companies use publicity to develop or improve the standing and goodwill of their brand in the marketplace.

11. **Promotional strategy:** The promotion mix is significantly influenced by the company's promotional strategy, specifically whether it employs a Marketing Approach or a Pull Strategy. In a "Push" method, the product's maker compels the dealers to stock the item and market it to the consumer, persuading potential customers to purchase it. Here, trade marketing and personal selling are probably more successful.

    When using a pull strategy, buyers urge retailers to stock the item; in other words, they make the purchase themselves. In this case, consumer marketing and advertising are more suitable.

12. **Other factors**: There are also smaller aspects that influence the promotional mix in addition to these main ones. These variables might include the product's pricing, the marketing channel used, the degree of product difference, the ambition to penetrate a market, etc. A promotional plan should only be developed after considering all relevant factors. The marketing manager has to comprehend these elements. These factors affect various organisations in varying degrees depending on the company's specific marketing environment.

3.3 Promotional Budget

A promotional budget is a set sum of money used to promote a business's or an organization's products or values. Budgets for promotions are created to take into consideration the essential costs associated with growing a business or maintaining a brand.

A marketing budget is an amount of money set aside for the marketing, marketing, or sales of a product or brand. The marketing budget for a brand-new or existing product will be established by business statistics, market research, and anticipated return on investment.

Spending on print, broadcast, tv, web, and other kinds of advertising is frequently included in budgets for promotions. A company's marketing budget may include the price of email campaigns, social media interaction, and outdoor signage.

Employing independent consultants and experts who develop campaigns and position adverts in the appropriate media and locations may also be done with promotional funds. Hiring marketing intelligence firms to analyse data that shows how annual budgetary dollars convert into new or recurring business for the organisation may be necessary to achieve this.

Businesses spend a lot of money on marketing initiatives. The budget for marketing and sale is determined using **four different techniques**. These are described in detail:

1. **Affordable method:** Businesses commonly utilise the cost-effective approach to determine the budget for marketing. What the firm can afford determines the promotion's budget. This method lacks consideration for a service's continuing promotional needs, making it a subjective assessment. This strategy ignores how promotions affect sales volume. Long-term planning is difficult since employing an economical technique typically results in an annual budget that is unexpected.
2. **Percentage of sales method:** This method determines the percentage of sales that go for promotions. The proportion of sales strategy has the following advantages:
    - First, there is indeed a direct link between advertising expenditures and sales. The company can therefore easily afford to spend a specific percentage of revenues on advertising.
    - Second, this strategy makes it simpler to examine the relationship between advertising costs and unit sale prices.
    - Third, even when other companies utilise the same percentage of income for advertising, this strategy ensures consistency.
3. **Competitive-parity method:** The company's advertising spending is comparable to that of its competitors under this technique. This method imitates the rivals' approach in terms of promotional spending. It is based on the premise that an investment made by a rival represents the industry's caution.

    A promotional war is avoided because the promotional spending of one firm is equivalent to that of the competitor. But this strategy has a number of shortcomings. There is no assurance that competitors' promotion budgets accurately reflect the collective wisdom of the sector. Organizations differ in terms of their standing, resources, objectives, and opportunities. Consequently, the marketing budget that works for one firm may not work for another.

4. **Objective and task method:** With this strategy, marketers create the advertising budget by detailing the objectives they hope to achieve, outlining the steps necessary to achieve them and assessing the expenses involved. This strategy makes sense since it puts the marketing budget at a level that is required to meet the company's objectives.

Promotional budgeting in pharmaceutical companies

Every pharmaceutical company has to decide the minimum and maximum level of promotional budget that they are going to spend. The economy of scale lies in the minimum level to fulfil the market requirement and maximum level is depending on amount of returns in terms of sales that company is expecting to get.

There are various methods of promotional budgeting depending on the size of the firm, objective of the company, market competitions. The important methods of promotional budgeting are discussed below:

1. **Fix percentage of sales method:** Here, a fix amount has been allocated as a promotional budget for the financial year depending on previous sale or projected sale.
2. **Fix amount allocation method:** As per this method company uses a fix allocated sum amount per product as per past sales and frames the strategy as per it.
3. **Economy of scale method:** Companies are using this method as top management wants to avoid such promotional cost that defines promotional budget at minimum level to get maximum output of it.
4. **Competitive market method:** Pharmaceutical companies now using this method frequently because of its own advantages like this promotional budgeting method is as per the movement of market and able to beat the competition in very specific market.
5. **Strategic objective and scheduling method:** This is the most popular method in pharmaceutical marketing from strategic angle that past sale is most important. The predefined objectives of sales are accountable in this method. The promotional budget allocated as per objectives and is scheduled as per planning of the product sale.
6. **Push-led strategy of product promotion:** Majority of local company does follow this strategy aggressively as product promotion, which is totally based on trade margin given to retailer. Generics without any brand power are being pushed into the market at huge margins to the trade.
7. **Goal-and-Task method:** This strategy is developed by setting clear goals, determining the steps needed to achieve these goals, and then estimating the costs involved in completing these tasks. This strategy is frequently employed to accomplish long-term objectives like increasing market share or brand recognition.

8. **Wallet method:** With this approach, monthly marketing campaigns are planned based on "what's available" as opposed to "what's the sales objective." Due to a lack of forethought, this strategy could limit revenue potential. But given that some businesses view marketing promotion as an expenditure rather than an investment, this strategy is often used.
9. **Co-op only method:** Using this planning technique, a budget is set out exclusively for manufacturer's cooperative (co-op) advertising assistance. This might be detrimental since the manufacturer's innovative message approach and the amount of co-op money that is accessible to the firm employing the co-op are both constraints.
10. **Zero method:** The marketing investment should be kept as low as feasible using this strategy. This approach is occasionally regrettable, especially when going out of business advertising effectively moves goods.

3.4 An Overview of Personal Selling

Personal selling is the practise of communicating directly with customers via paid representatives of an organisation or their agents in a way that makes the audience believe the communicator is speaking on behalf of the business.

The method of communication used by salespeople to develop a personal connection with clients and provide value for the company. Sales for the sales team and advantages for the client may be the value. The value could include giving customers knowledge as well as cash advantages. For instance, CIPLA pharmaceuticals' medical representatives inform doctors but do not attempt to sell them drugs.

Approaches to personal selling:

- **Stimulus response selling**: In this scenario, the salesperson presents the stimulus and anticipates the customer's reaction. Until a purchasing decision has been made, this procedure will continue.
- **Need satisfaction selling:** The sales representative determines and validates the customer's need for the goods. He offers the consumer a variety of options from which to pick, and the process continues until the buyer makes a purchase.
- **Mental states selling**: It includes: get action by attracting attention, holding interest, generating desire. In actuality, the majority of seasoned marketers concur that when preparing their presentation, they do not consider the numerous mental states.

3.4.1 Problem Solving Selling

When a consumer has a purchase issue, this strategy is applied. The sales representative defines the customer's issue in this method, develops potential solutions, and assesses them. He continues to work on a specific solution until the customer makes a purchase.

3.4.2 Methods and Characteristics of Personal Selling

Personal Selling has several methods with their characteristics properties which have been mentioned below in **Table 3.1.**

Table 3.1 Personal selling methods and characteristics

S. No	Methods	Characteristics
1.	***Retail selling*** the is selling to ultimate consumer.	Personal selling is flexible.
2.	***Field Selling*** is business-to-business selling that occurs in the place of business of the potential client.	Personal selling builds relationships.
3.	**Telemarketing** is by primarily engaging with potential clients over the phone, and telemarketers frequently utilised computers for order taking.	Personal selling produces immediate response
4	**Inside Selling** is the location of business where the salesperson works a business-to-business transaction	

The conditions favorable for personal selling is either when a product is in the introduction stage of its life cycle or when its price is high, it is highly technical, and it requires demonstration. One instance of personal selling is when an organisation lacks the funds to carry out advertising campaigns. Personal selling is advantageous if the goods can be personalised and the market is small.

3.4.3 Steps involved in Personal Selling

1. **Identifying the prospective buyer:** Future customers are identified in this phase of the personal selling process. It's possible that not all recognised prospects will really end up as clients. Finding the right prospect is essential since it influences how future sales will be made. Marketers look in many different places to find prospective customers. Prospects are sought for by marketers using directories, the internet, phone calls, and postal mail.

   At trade shows and exhibitions, marketers put up a booth, gather contact information from current customers, and cultivate referral sources like vendors, dealers, salesmen, executives, bankers, etc. After identifying potential customers, the salesman evaluates them for suitability based on their needs, wants, preferences, and tastes.

2. **Pre-approach:** The step after qualification and prospecting is pre-approach. The salesman now has to decide how to contact the potential customer. The salesman could personally visit them, give them a call, or write them a letter depending on how convenient it is for the potential clients.

3. **Approach:** At this time, the salesperson should appropriately approach the prospects. The salesperson needs to formally introduce themselves and start the conversation. The salesperson's mood, manner, and speaking style are currently the most crucial factors.
4. **Presentation and demonstration:** The salesperson now product description and its benefits in great detail. The salesperson outlines the benefits of the product, discusses its features, and estimates its cost. It is important for the presentation to be clear and enticing.
5. **Overcoming objections:** After a briefing and presentation, clients are asked to make a purchase, but many are reluctant to do so and raise concerns. Customers prioritise well-known brands and show signs of indifference, impatience, reluctance to converse, among other things. The client may take issue with the cost, the timing of the delivery, the characteristics of the item or the company, etc. The salesman expertly addresses the buyer's objections by outlining them, and ultimately convinces the consumer to purchase.
6. **Closing:** The salesperson requests an order from the client after addressing their concerns and convincing them to buy the products. An order is placed with the assistance of a salesperson.
7. **Follow-Up and maintenance:** After resolving their worries and convincing them to buy the product, the salesperson requests an order from the customer. When a consumer places an order, the salesman assists them.

   Personal selling works well for freshly released products and products that need to be demonstrated and presented.

Personal selling and Pharma industry: Personal selling is one of the most traditional and successful types of company marketing. It is commonly required yet expensive due to rising customer and product sophistication as well as greater competition. In a highly competitive industry like the pharmaceutical industry, where many businesses sell identical or similar ('me-too') products, the distinction is typically made by the individual behind the product.

Selling is indeed the single most important promotional approach used in the pharmaceutical industry, which has led to a unique situation regarding the sector's marketing focus. Being in continual interaction with clients, markets, and competitor activity puts a salesman in the pharmaceutical business in the ideal position to execute the marketing orientation.

3.4.1.1 Role and Responsibilities of Salespersons in Pharma Industry

Salesperson in Pharma industry is commonly known as 'Medical Representative' who informs physicians about the company's goods provides their accessibility to the market and is accountable for ensuring prompt payment from the marketing middlemen.

A medical representative's duties include client service, prescription generation, and market research.

Prescription generation: The most important task for a medical representative is to increase sales, and the best method to achieve this is to get doctors to write prescriptions for the company's goods.

A doctor's written prescription instructs a patient on which medication to take, how much to take, how often to take it, and for how long. The patient purchases the required prescriptions at the pharmacy after bringing this notification with them. The process of writing a prescription is initiated when the patient pays the pharmacy (or "retailer") for the medications from the suggested product (or "brand"). As a result, writing prescriptions for the goods is the sole factor (or "pull") that causes pharmacists to stock and sell the medicines while also producing money for the business. This is a challenging role that calls for serious efforts, frequent visits to the doctor and pharmacist or stockist, correct use of advertising material, such as books and pharmaceutical samples, and rigid adherence to the work rules established by the firm.

Customer coverage: The standard visiting list, or SVL, is a list of doctors that the medical representation must regularly communicate with in accordance with the established "daily work plan" and to whom they must deliver the essential medico-marketing material.

Medical representatives must be knowledgeable of the various specialisations of doctors in order to offer the appropriate items to the appropriate physicians. The medical representative is responsible for gathering feedback on the goods that the physicians use and for properly resolving any product-related complaints or problems, as needed with assistance from head office. The responsibility of the medical representative is to meet with the pharmacists in a certain region to assure free access of the goods. He can also personally take orders from pharmacies, transmit them to stockists and distributors, and follow up on them if necessary. Products that are about to expire, breakages or leaks during delivery, and credit or debit notes are only a few of the problems that frequently demand swift and satisfying resolution with direction from superiors when necessary. One has to ensure that each and every call made on a customer result in a productive output.

It goes without saying that the only way to accomplish the intended outcomes for both the business and oneself is to build a solid working connection with one's clients that is built on mutual trust and respect.

Market intelligence: In light of the efforts of its competitors and the situation of the market, a company's ability to satisfy the requirements of its consumers determines whether it succeeds or fails. The only person qualified to provide this information is the medical representative. The right customers to market to, the right goods to promote to them, and trying to thwart competitors' tries to target the very same clients for their goods by promptly activating superiors are all crucial parts of a representative's job description. This allows for the creation of the best marketing strategy.

When done at the chemist level, a "prescribing audit" or "branding audit" is an extremely useful technique for gathering vital information regarding doctors' prescription practises and pharmacist stocking practices patterns, slow- and fast-moving products, and promotional activities carried out by rivals.

3.5 Advertising

Advertising By describing the product's features, applications, cost, and source of availability, a strategy is utilised to make the public more familiar with the item. Advertising has many functions, such as educating consumers about a product, convincing them that a company's products or services are indeed the best, enhancing the professional image, trying to identify and generating demand for products or services, showcasing new uses for current products, announcing new products and services, continuing to support salespeople's individual messages, attracting customers to a business, and retaining existing customers.

The advertising message must be heard by a billion people who speak various languages and follow numerous religions. Advertisers may contact their target audiences through television, radio, film, newspapers, outdoor media, sales promotion, as well as the Internet. As a result, mass communication is a sort of marketing. As seen in **Table 3.2**, advertising has advantages and disadvantages of its own.

Table 3.2 Merit and demerits of advertising

S.No	Merits	Demerits
1.	Supports new product in market and enables a company to face competition in market.	Escalates the cost of products
2.	Promotes sales of goods and services by enabling a company to improve its reputation.	Increases people's demands by encouraging them to purchase more things
3.	Stimulates the demand of product and provides information to customer about product and prices.	Advertisements sometimes give misinterpreted information's as in Beauty creams.

Steps for development and execution of advertising:

(i) **Briefing:** The briefing, which is an experience and competence of the client and marketing agency's grasp of the product to be marketed, the advertising purpose, the timeline for the program, the ways to reach the target, and the total anticipated cost, marks the start of the advertising process.

(ii) **Market research:** Market research will begin after the briefing. Comparing an advertiser's product or service to that of a competitor, gauging how consumers view their brand in relation to that of competitors, examining competitors' advertising, and gauging how consumers react to that advertising are all examples of research.

(iii) **Identify target audience**: The next stage is to determine your target market. The advertising firm will determine the target market using the market research.

(iv) **Media selection:** In order to reach the desired population in the most cost-effective way, the advertising company or media agency will select the medium depending on the study.

(v) **Designing and creation:** The creative team at the advertising firm will now translate the marketing communication into text and images. A copywriter will create the advertising copy, while an art director will use images to visually represent the copy. The advertising agency may contract with other production firms to handle the filming or taping.

(vi) **Decide place and time:** Choosing where and when to display the advertising is the next step. The advertising agency's traffic department will make sure the advertisements are ready on time and have received the necessary legal permissions.

(vii) **Execution:** The advertising will thereafter be put into action.

(viii) **Performance evaluation:** After the advertising is run, the news outlet will assess its effectiveness.

Importance of Advertising: Advertising is extremely important in today's atmosphere of intense competition. Advertising is such thing that has grown crucial for everyone in today's daily lives, whether they are manufacturers, traders, or customers. Advertising is an important element. Let's analyse the importance of advertising:

1. **Advertising is important for the customers:** Imagine it. Without an advertising on the radio, television, or newspaper! No one can ever conceive this, I assure you. Customer life is greatly impacted by advertising. Customers are individuals who buy products after learning about their possibilities in the marketplace. Regardless of whether the product was advantageous to them, no customer would discover the products without promotion, and they'd not buy the goods. Advertising also helps people choose the best products for themselves, their families, and their kids. Consumers may compare products when they are aware of the range available and make choices to guarantee they get what they want after spending their hard-earned money. Advertising is crucial for customers as a result.

2. **Advertising is important for the seller and companies producing the products because of the following reasons:**
   - **Advertising contributes to rising sales.**
   - Manufacturers and enterprises may identify their competitors through advertising and develop competitive strategies.
   - Any business intending to debut or promote a new product into the marketplace needs a solid basis, which advertising provides. In order for people to visit and test the new product, advertising helps to raise awareness of it.
   - When a consumer reaches a specific age, advertising helps the company develop brand loyalty.
   - With the assistance of advertising, the cycle of production and demand for the product keeps expanding.
3. **Advertising is important for the society:** Advertising educates consumers. Advertising also addresses various social concerns, such as child labour, alcohol use, the murdering of young girls, smoking, contraception education, etc. Given this, advertising is important for society.

   Potential media utilized in advertising gives information of the product through most common and one of the most effective medium for advertising as **Television**. Advertising in distant located and less developed village areas given by radio. **Newspaper, magazines, posters, balloon advertising and banners gives** advertising of each and every daily need product (including pharma products).

   For pharmaceutical advertising, one's decision of purchase is not made by the patients (actual consumer) but by the doctor (intermediate customer) who prescribes the drugs to patient. Hence, this is an example of indirect advertisement. So the advertising method should be design and formulate so as to generate more prescription of the product from the doctor. The Special methods used for advertising and sales promotion in pharma industries are:

   Free samples: Samples are distributed to doctors in order to help them to conduct free trail of drugs. In western countries these are usually distributed at the time of introduction of new drug in market but in India sample distribution is a regular feature.

   Prescription pads: The prescription pads of various sizes, shapes, designs and printing are provided in companies to the doctor. However, doctors mostly use them for writing other things.

   Advertising in journals: The advertisement of prescription drugs in public media is not allowed in India by the enactment of "*The objectionable advertisements and magic remedies act*" however the drugs can be advertised in scientific journals and distributed to the medical professionals, hospitals or a medical/pharmaceutical laboratory example, **CIMS** (Current index of medical specialties), **MIMS** (Monthly index of medical specialties) and **Drug today and Drug update.**

Calendars and diaries: These are the common features during January every year. They are printed by pharmaceutical companies in various sizes, designs and utilities along with the name and product of company printed on them and given to the doctors and other medical professionals.

Sponsorship programs and conferences: Many pharmaceutical companies sponsor various medical conferences and workshops for this the companies pay a fund amount to the organizers and in return they get the space for exhibition of drugs, literature, banners, machinery or audiovisual programs related to drugs. These types of activities enable to create a mass contact with doctors.

Direct mailing: In this strategy, the corporation compiles a list of RMPs (Registered Medical Practitioners) and pharmacists, and then sends letters, pamphlets, and catalogues to these individuals to tell them of the most recent information regarding the medications produced by the company.

Gifts: The practice of providing gifts to doctors is becoming a popular aid in advertising of pharmaceuticals day by day. In most of the cases the name of product and company is printed on the gift. Thus they reminder the doctor about the company and its drug.

Development of advertisement program: The development of advertisement program involves following three basic steps:

1. **Determine the advertisement objectives**: In this initial phase of advertising the company should determine the major objectives of advertisement. There can be one or several objectives of advertising as mentioned: To spread the word about the product, show how different it is from those of the competition, and boost sales of the product in the final stages of its life cycle.
2. **Determine the budget of advertisement:** The total budget that has to be utilized in advertising depends on several factors-
   (a) **The corporation seeks to spend greater money on marketing** during the first stage of the product life cycle, which is the period in which the product exists, to raise awareness of the product.
   (b) **Market competition** faced by any product which has to face high competition in market due to availability of large number of already established products, the company has to spend more budget on advertising. It helps to build uniqueness and differentiation of product image in market.
   (c) **Available resources** usually larger companies can spend more money on advertising due to better availability of funds.

(d) **Number of products launched** by a company, as launching single product at a time in market, tends to utilize more economic resources on advertising this product because company's profits are solely depending on the sale of this product only.

(e) **Selection of advertising media** for targeting the customers and their interests, the advertising media should be chosen. The selected media must be effective and affordable within the allocated advertising budget.

(f) **Training of advertising personnel's (Medical representatives)** is very important for any company as the trained medical representatives are in the distribution system of India for providing information of the pharmaceutical product. The training is helpful in imparting knowledge about the Anatomy/Physiology of human body, Pharmacology of concerned products, advantages/ disadvantages, features, benefits, side effects and dose related information of drug.

3. **Evaluation of advertising techniques**: It determines whether advertisement message is inducing the doctors to prescribe the drug and encouraging consumers to use products (In case of OTC drugs or cough tablets/ lozenges). This evaluation can be done by simple methods like:

   (a) **Method of recognition**: Showing the commercial to the public or medical professionals and asking them if they have ever seen it.

   (b) **Aided Recall**: Invoking the public to identify the brand of medication used to treat a specific illness.

   (c) **Unaided recall**: Inquiring about the respondents' memories of viewing any advertisements for a certain product category.

3.6 Direct Mail

By delivering printed mailers, dimensional packages, perishable commodities, promotional items, or other tangibles, marketing professionals use direct mail marketing as a strategy to contact with potential customers and consumers in person. It is used by marketers of all hues, including recruiters, marketing teams, demand generation teams, development teams, and outbound and inbound sales teams, among others. Direct mail marketing is a successful tactic for engaging with customers, leads, and prospects. It's an initiative that sends a printed advertisement to a specific client through a mail service.

Direct mail marketing as a useful component which promotes overall marketing campaign by the following ways:

- **Making direct mail more memorable:** The human touch is likely what makes direct mail most memorable; a handwritten message or a present that was individually chosen can forge stronger emotional ties than a generic corporate gift.

- **Direct mail has a higher response rate:** Email is the most effective digital medium, but direct mail marketing has a response rate that is 30 times greater due to this tactile contact.
- **Less direct mail competition:** Since so few businesses allocate money to tactile marketing, direct mail is especially noted for its scarcity. A direct mail marketing item stands out from the competition because of this.
- **Direct mail offers many ways to be creative:** Direct mail offers countless opportunities to surprise and amuse recipients. Marketers may send a wide range of artistically presented things through a shipping platform, including handmade cards, cutting-edge swag items, personalised presents, candies & snacks, e-gift cards, event tickets, and more.
- Directly address the target audience so that you may learn who receives the message, when it is delivered, what is inside the envelope, and how many people it touches.
- It is highly targeted, measurable, individualized and cost effective.

3.6.1 Direct Mail for Pharmaceutical Companies

Direct mail has long been used by pharmaceutical companies to educate physicians and consumers about the benefits of their products. With something as important as healthcare information, direct mail and printed communications in general have been great tools for driving engagement and conveying all the important details of a company's new drug. Various tools to be addressed by pharmaceutical companies while promotion of a product through direct mail, such as:

1. **Use of various paper weights and sizes:** Doctors' clinics receive a significant amount of mail in plain white letters, so if a product is to stand out, one must consider employing lively, tactile marketing materials like heavy, glossy postcards or coloured paper.
2. **Make use of data and reviews:** To make the mailers specific to a particular practise, genuine statistics may be used, and they will inform the physicians about patient testimonies.
3. **Be genuine and specific:** The mailer has to be distinctive because doctors are among the groups that is more intensively advertised. As an illustration, think about using accurate data and locating patient testimonies that demonstrate how well they are feeling. The phrase "nearly 85% of our clients are capable of returning to work" is a good example of a verifiable assertion. This puts what physicians can do by sending your patients into real-world context and context!
4. **Personalizing pharma Direct Mail:** The pharma direct mail can be personalized by the health care professionals (HCP) by addressing their correct name, title, clinical setup.

3.7 Journals

It is a written/printed material that deals with a particular subject (product/service) or professional activity. A journal of marketing advances and disseminates information to academics, educators, managers, customers, policy makers, and other social stakeholders concerning goods and services or other professional activity. While most experts agree that medical journal advertising has been underutilised in the past, pharmaceutical marketers can benefit from it.

Professional medical publications routinely outperform other sources of information for doctors and are the most favoured and reliable ones.

The widespread consensus is that peer-reviewed medical publications are a good place to find neutral, trustworthy information regarding medications.

Drug corporations advertised in medical magazines for $448 million in 2003. Pharmaceutical marketers hold medical newspaper advertising in high regard since it may quickly raise awareness because to its cost-effective reach and regularity.

Pharmaceutical corporations may get prescribers via medical publications, which can provide a range of strategies to target specific prescribers. Various advertisements are distributed to different physicians as a consequence of financial agreements between journals and pharmaceutical corporations.

Journal advertisements generate profits not only for pharmaceutical companies and medical journals, but also for the physician organizations that publish the journals. Pharmaceutical industry advertisements in affiliated medical publications contributed 10% or more of the total yearly revenue to five of the six physician associations in 1996; the percentage of total revenues varied from 2.1% to 31.3%. 15.1% of the American Medical Association's (AMA) overall revenue in 2004 came from advertising in AMA publications. The AMA collected $40.7 million in revenue from magazine advertising. The New England Journal of Medicine, Annals of Internal Medicine, Medical Journal of Australia, Indian Journal of Medical Research, Indian Journal of Community medicine, Indian Journal of Critical Care Medicine, and others are a few examples of periodicals.

Sampling:

The consumer receives a free sample of the product so they may check it out before making a purchase. A piece of food or another product (such as beauty items) provided to customers at shopping centres, supermarkets, retail establishments, or via other channels is known as a free sample or "freebie" (such as via the Internet). The consumer may make an informed choice regarding a product's qualities after experiencing its quality and value through sampling.

Free samples are now widely available on many consumer product businesses' websites in an effort to entice customers to use the products frequently. One of the most crucial sales marketing strategies is the use of samples. Samples are described as offering of a limited quantity of a product to consumers for testing. Customers receive free samples to pique their fascination with the company. Samples allow customers to check the product's quality.

Consumers receive samples at their doors. Free coupons that may be redeemed at retailers or distributed to consumers inside the physical store are also provided to customers by mail. Samples are occasionally affixed to other products. Even while sampling is useful, it is expensive to produce several samples of a given product. Additionally, giving out samples to customers costs money.

Effectiveness of sampling marketing: As shown in **Figure 3.2**, there are four main ways that product sampling can benefit a business.

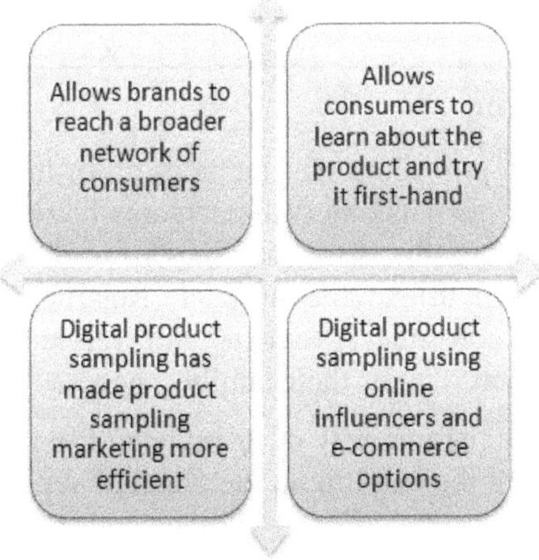

Figure 3.2 Effective parameters of sampling marketing

In situations where the product is an established one, **sampling** is not justified. A product won't be eligible for sample if it isn't in some manner better to rival items or has a low turnover rate. Additionally, sampling is not appropriate for perishable, bulky, or products with a small profit margin.

Why product sampling marketing works? Pre-purchase product testing has been a major factor in rising sales and improving brand recognition.

The use of digital platform options for product sampling marketing has significantly increased sales conversions. According to a research by Edison Media Research, 35% of consumers who try a sample before making a purchase will do it on the same trip to the store.

What Makes Product Sampling So Effective? The psychology of sampling is straightforward: people love getting free things. Giving up free samples encourages customers to trust the company, which increases the likelihood that they will have a favourable opinion of it.

Other elements, such as certain significant consumer insights, also have a role in its efficacy:

(a) **Less risk aversion:** Trying out a product doesn't need the recipient to make any investments or commitments. After using the sample, the consumer may be less unsure about a brand's performance, according to behavioural changes in the testing of new items. Sampling may be a subtle strategy to influence consumers to pick one brand over another.

(b) **Greater reciprocity:** When a merchant or shopkeeper receives a free sample, they are more likely to repay the favour by placing an order in the future.

(c) **Increased brand trust:** Giving customers the chance to hold the goods in their hands for themselves forges an emotional connection between them and the brand.

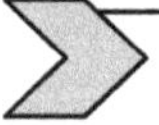

3.8 Retailing

Retailing is a marketing procedure that includes all of the actions necessary to sell the goods directly to the customer (the person who will utilise the product). It includes the selling of products and services from the a point of sale to the final consumer who will utilise the product. Retailing includes all actions necessary to promote consumer products and services to final customers making purchases for their own or their families' needs.

Any firm that derives the majority of its revenue from retailing is referred to as a retailer. Retailers are the last link in the chain of supply between producers and the final customer.

Characteristics of retailing

(a) **Direct customer contact:** The retailer serves as the last point of contact between a firm and a client. The retailer recognises the customer's need and offers the best remedy.

(b) **Little-quantity purchases:** Customers only buy a small amount of goods from the retailer. Despite the fact that the consumer bought less than usual, he nonetheless made a transaction. Customer and store relationships are improved as a result.

(c) **Customer service:** A store offers its consumers a variety of services, including home delivery and credit options..

(d) **The significance of location and design:** Businesses place point-of-purchase displays at retailers. Layout is just as crucial as location, especially in the modern retail industry. Ambience is a need rather than a luxury when it comes to the extended marketing mix. Additionally, they urge retailers to use word-of-mouth advertising to market the goods.

(e) **Employment provider:** The second-largest employment in India is the retail industry.

(f) **Marketing communication:** Businesses place point-of-purchase displays in retailers. They further urge merchants to use word-of-mouth advertising to market the goods.

Functions of retailing:

(a) Provide outlets for products from manufacturers.

(b) Retailers place the products in the right sequence so that customers may quickly recognise the products or services.

(c) Retailers are located in areas convenient for consumers.

(d) Keep enough stock of products on hand to satisfy customers' daily needs.

(e) They provide advice and after sales services to consumers.

(f) Provide useful feedback from consumers to wholesalers or manufacturers.

(g) Retailers use word-of-mouth marketing to spread the company's merchandise.

(h) They provide consumers with the products and services they want in the quantities they choose.

(i) In the case of durable products, the retailer agrees to door delivery orders..

Types of retailing:

A. **Store retailing**: the type of retailing where having a store in a certain area is necessary for conducting business. It can be presented in a variety of formats:

(a) **Speciality store**: The shops sell a lot of goods, but only in a few product categories, like furniture or textiles. Take Tanishq, a retail jeweler, as an illustration.

(b) **Departmental store**: Clothing, home furnishings, consumable products, and services are all sold in this retail style. Every one of the forms is handled as a separate department in the retail establishment. Example: Raheja Group customers cease.

(c) **Supermarket**: It is a self-service business that is comparatively big, low cost, small poor, high volume, and geared to meet all of the consumer's demands for food and household goods. Examples are Big Bazar and Reliance Fresh.

(d) **Convenience store**: Customers' homes are extremely near to these stores. It often sells high-demand items with a high turnover rate at a premium price. For instance, Reliance Fresh.

(e) **Discount store**: These shops provide things for cheap with low profit margins. The retailers produce large volumes in order to make a profit. This style is used by Subhiksha, a retailer in south India.

(f) **Off price retailer**: This kind of retailer purchases the products for less than wholesale. These goods are sold for less than the suggested retail price.

(g) **Super stores**: These are relatively sizable establishments where clients may buy both food and non-food items. The superstore has category killers, which carry a lot of goods in a specific category. An such is Nalli Sarees, which has a wide selection of sarees in its outlets.

(h) **Hypermarket**: These stores sell large items and have spacious interiors. Reliance Mart, for instance, in Ahmedabad.

B. **Non store retailing**: the method of retailing when a business sells its goods through electronic or direct selling channels. As an illustration, consider direct marketing, telemarketing, automatic vending, internet shopping, and direct marketing.

Retailing in Pharmaceutical industry: Contrary to hospital pharmacies, which are often referred to as community pharmacies, **retail pharmacies** sell medications directly to patients.

Any independent pharmacy, chain pharmacy, supermarket pharmacy, or mass merchant pharmacy that has a state-issued pharmacy license and sells prescription drugs to the wider populace at retail pricing is referred to as a "retail community pharmacy."

Retail pharmacists advise the public on basic healthcare issues and provide both prescription and over-the-counter medications. The essential qualities of a retail pharmacist include responsibility, maturity, and attention to detail. Retail pharmacists must have an excellent interpersonal skill and organizational skills, with good verbal communication, and have confidence and have awareness of the market. There is a proper mechanism to handle a retail pharmacy by the retail pharmacist as mentioned in **Figure 3.3.**

Role/ Functions of a retail pharmacist:

- Responding to consumer inquiries with advise on medical conditions, symptoms, and drugs.
- Hiring, educating, and supervising employees.
- Handling prescription processing and drug distribution.
- Purchasing and reselling supplies like pharmaceuticals.
- Speaking with medical professionals.
- Offers marketing-related services.
- Handling financial management.

- Keeping and maintaining financial and statistical records.
- Creating displays and promotional materials.

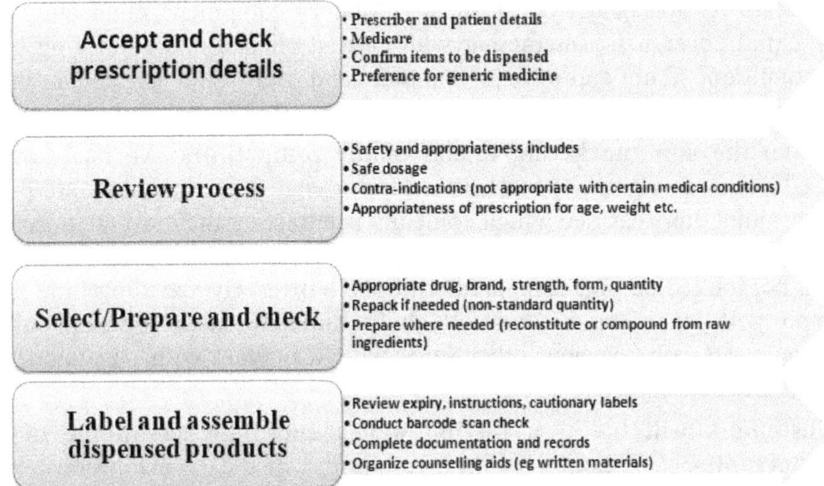

Figure 3.3 Mechanism of Retail Pharmacy

3.9 Medical Exhibition

A medical exhibition is an event where the medical goods (medicines/surgical equipment's etc.) and services of a specific company are exhibited and demonstrated to public. In a medical exhibition or a medical expo provides the platform for world-class medical equipment's for sale. Visitors can find a one-stop store for buying medical equipment's for their hospitals or clinic, get dealership from international companies and more.

General benefits:

- Medical exhibitions are available to a wide and occasionally diversified range of client demographics, and they serve as a platform for the promotion of healthcare and biopharmaceutical products to the target audience and the development of brand recognition (usually the general public). This gives you a chance to spread the word about your goods and services to a larger audience that might be unaware of them already.
- You may utilise an exhibition to learn what the market or the general public thinks of your product. Through the visits to your display booth, you have the chance to expand into business-to-business trade and build a client database.

Risks: Requires enough fund and manpower and time.

Advantages of Medical Exhibition

1. **Visibility to acceptability, exhibiting at medical has hundreds of benefits for the business.** Medical trade exhibitions provide a strong platform for releasing new medical laboratory equipment, connecting with potential consumers, interacting with current clients, and enhancing brand recognition. With appropriate planning and right strategy, exhibiting at medical can be a productive and cost-effective choice.
2. **Learn the new marketing tactics of the competitors:** Medical hospital needs expo as a great opportunity to learn what the competitors are doing. The competitors can see which company is attracting the most attendees and can learn from them. The prices and special deals offered by the competitors can be; looked on and compared with one's offer. By spending time in the expo will make one understand about the new sales strategies of the competitors and see how other sales people interact with customers, and what are their objectives.
3. **Customers will be in a buying mood and take advantage of this opportunity:** Without feeling resentful, medical exhibition offers businesses exceptional access to key prospects. Customers will be in a buying mood and you have to take advantage on this opportunity. There are chances that the attendees are already focused for the products or services, so, take a direct sales approach and view every interaction as a chance to close the deal. Try to schedule meetings with the important prospects for a week following the healthcare trade show if one is unable to seal transactions. A medical device trade exhibition is a chance to significantly increase the consumer base for your business if you approach it strategically.
4. **Develop and strengthen your brand:** A key component of company success is branding, so develop an experience that will let booth visitors connect with and recall your brand on a deeper level. Engage with every visitor because they could become clients. To establish a strong brand identity in the eyes of your clients, use eye-catching, identifiable display banners and locate your booth close to the leading businesses in the sector. Exhibition of Medical device trade show can transform small start-up business into a profitable enterprise.
5. **Launch of new healthcare product into market through medical exhibition:** One of the finest justifications for an exhibition is the introduction of a new piece of medical equipment or healthcare product. The chance to find cutting-edge new medical items is among the most exciting trade fair opportunities for guests in the medical business. A successful product launch may increase early sales and attention in a product.

   Benefits of launching the new healthcare product in medical exhibition: The medical exhibition continually rising popularity due to what it brings to exhibitors and prospects. In the current healthcare industry, it is important than ever to build and maintain trust with the clients. Even though the advances in technology insist better ways to market a product, nothing can replace the value of face-to-face marketing. Medical exhibition helps in productive marketing and the benefits have been summarized below:

- **Interested attendees:** Medical exhibition gives exhibitors the chance to market their new offerings to a large number of interested attendees. There are attendees who are purposively attending exhibition to see and buy new products. The customers are already interested in what they have to offer and are all in one big room together. It's like a marketer's dream come true!
- **Maximize initial sales:** Early stages of a product's development are frequently crucial to the ultimate success or the company's failure or product. A medical device trade fair gives an excellent potential for successful outcomes when a brand-new healthcare product. Potential clients are more inclined to follow, which will increase the sales of your goods. A new product needs medical support to gain the first following it needs to catch the public's attention. Use of a medical show as the launch party for a product builds anticipation for what will be revealed.
- **Surveys and review:** The simplest method to get a tonne of evaluations and comments on a new product is to introduce it at a medical fair. It's the finest approach to get crucial information on what the general public thinks of the business offering directly from them.
- **Use of product:** Exhibitors can boost their lead generation results by giving attendees a more interactive experience. Consumers adores free sample products. The live demos help the customers to communicate the benefits of the product and getting new products into shopping carts. This is an opportunity to connect with the potential customers and gain feedback. A medical device trade exhibition offers the chance to reach a live, captive market and proves to be the best option to get your item in front of your intended market.
- **Saves time and money:** It may be highly expensive to broadcast commercials, and online marketing takes time to achieve a significant audience. A medical expo offers the chance to connect with a sizable crowd gathered in one location at the same time and to spread the sales message.
- **Create instant visibility:** This is a significant distinction between marketing at medical expos and other outlets for advertising. Attendance at medical expos demonstrates support again for healthcare sector. In fact, it has been discovered that although some visitors may not make purchases at the expo, they are far more likely to choose a vendor that participated in the trade show when they are ready to make a purchase.
- **Builds trust between customer and consumer as a profitable relationship**: In a medical exhibition, kinds of interactions help to know about customers on an individual level and are very crucial for building trust

and establishing strong profitable business relationships. After the meeting, one can follow-up with through emails or phone calls because a relationship has been established.

3.10 Public Relations

Managing the information flow between a person or an organisation and the public is known as **public relations (PR)**. Continuous efforts to maintain a positive public image and the general health of the business are referred to as public relations. One of the key foundations for the Indian pharmaceutical industry's pharmaceutical marketing is this relationship. In the case of a health care system, such public relations are useful in preserving the connection with the end user, who is a health care professional.

Educating the public about a firm and its goods is another activity that falls under the umbrella of public relations. Public relations activities are frequently carried out through the media, such as newspaper, television, magazines, etc. As was already said, one of the main actions involved in promotions is frequently thought to be public relations.

Public relations is the management of an organization's internal and external communications to establish and uphold a favourable reputation. To gain the public's trust in their businesses, businesses nowadays have begun to establish specialised public relations departments. There are times when health programmes like the national pharmacy week, the vaccination programme, the national diabetes week, and the pulse polio campaign can be funded. These programmes support the development of a professional image and customer engagement. The public relation plan has following steps as mentioned in **Figure 3.4.**

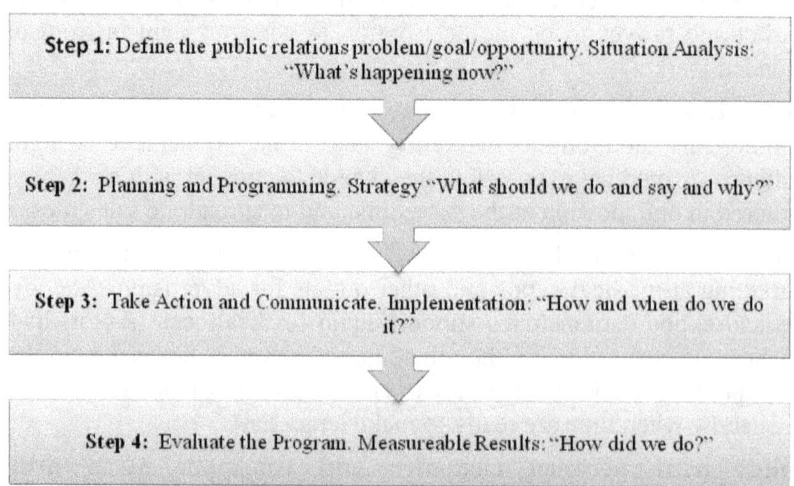

Figure 3.4 Steps of public relation plan

Any effective market penetration is facilitated by a comprehensive PR approach. This will increase consumer awareness of the brand/product and forge a solid position for the brand in the market. The interested parties are kept informed and are able to monitor consumer reaction to the product.

Methods of communication through Public Relations:

Newsletters: A good way to market a business, interact with customers, and inform them of new goods and services is to publish or send out newsletters.

Social media: Bypassing the media and reaching clients directly will be made possible by this. By following and being followed by journalists on social networking sites like Facebook and Twitter, you may enhance visibility for your company brand, drive web traffic, manage issues by promptly reacting to critiques or poor views, and handle difficulties.

Business events: These occasions offer businesspeople the chance to advertise their companies, market fresh goods and services, and make sure that pertinent information reaches the right clients.

Buzz marketing: Buzz marketing is a type of viral marketing that focuses on maximising a campaign's or product's potential for word-of-mouth promotion, whether through small-group talks among customers' loved ones or more extensive debates on social media sites.

Others: To get media attention and clients, lobbying organisations hold news conferences or grand openings. Public relations are improved by using textual and audiovisual media to communicate with the public. Public service initiatives have demonstrated the organization's social obligations.

Advantages of public relations:

(i) Assists in establishing and sustaining connections with the neighbourhood. For instance, Coca-Cola India's mission to change villages allowed the company to improve its reputation among rural consumers.

(ii) The PR maintains stronger ties with the investors.

(iii) Develop a positive reputation among social groups to generate word-of-mouth publicity.

(iv) It lessens disputes and misunderstandings regarding the business or the goods.

(v) Finally, aids in product promotion.

Disadvantages of public relations:

(i) Benefits of PR are difficult to measure.

(ii) Absence of control

(iii) Effective management is necessary.

Functions of public relations department: Public relations establishes a company's relationship with the general public. If the word is strong, it will spread across many mediums. When a problem arose, the corporation turned to public affairs rather than advertising. The following list outlines the duties of PR:

(i) **Press relations**, sometimes known as press agency, is the process of creating and distributing noteworthy material to draw attention to a subject, an item, or a service.

(ii) **Product publicity** entails promoting certain items and fostering relationships with the local or national population.

(iii) **Lobbying** entails establishing and maintaining connections with lawmakers and other government figures in order to have an impact on legislation and regulations.

(iv) **Maintaining contacts** with investors and other members of the financial community is a part of investor relations.

(v) In order to obtain **financial or voluntary support**, development requires public relationships with donors or users of nonprofits.

3. 11 Online Promotional Techniques for OTC Products

In contrast to prescription pharmaceuticals, which may only be purchased by customers with a valid prescription, **over-the-counter (OTC)** drugs are medications that can be supplied directly to a consumer without even a prescription from a healthcare practitioner. OTC medications are chosen by a regulatory body in several nations so that their active components are secure and efficient when taken without a doctor's supervision. Active pharmaceutical ingredients (APIs), not finished goods, are often what regulate over-the-counter (OTC) medications. Governments provide manufacturers the ability to combine chemicals, or combinations of substances, into unique mixes by regulating APIs rather than particular medicinal formulations. Launching an OTC product has a few fundamental benefits and drawbacks, which are shown in **Table 3.3.**

Table 3.3 Advantages and disadvantages in launching an OTC product

S.No.	Advantages of launching an OTC product	Disadvantages of launching an OTC product
1.	No restrictions on advertisement.	The speculative nature of the dealings causes market integrity issues.
2.	Marketing cost per people is less.	Specific nature of risk and scope is unknown to the regulators which increases the universal risk.
3.	Generate direct demand for customer.	Lack of transparency.
4.	Less dependence on doctors.	Lack of a clearing corporation or exchange, which increases the default risk or credit risk attached to each OTC transaction.
5.	Relatively less expenditure on sales force.	

Online Promotional Techniques: There are several promotional techniques required for OTC drugs:

Search engines: The practise of building a business website and blog to showcase the items in order to rank it better on important search engines when web users look for certain keyword phrases is known as search engine optimization (SEO). To create the internet presence and optimise the website for display, an expert is recruited.

Link exchange: Particularly if the marketing budget is constrained, it is done (or nonexistent). It entails looking for businesses that provide related goods and approaching them with an offer to swap text links or advertisements.

E-mail marketing: *E-mail marketing* is an online version of direct mail. Instead of sending a customer a flyer or advertisement, this form of internet marketing allows business to send the same info, or even more, *via* email. Business have a wide variety of choices on what to send in an email marketing campaign, including coupons, newsletters, invitations to special events and surveys.

Pay-per-click advertising: Another name for this is search engine marketing. Pay-per-click marketing is when companies display adverts on apps like YouTube as well as web search websites like Google. The majority of search engines provide companies the option to bid for advertisement space.

Article marketing: One may use the resource box at the conclusion of informative articles to draw readers to the website by authoring and submitting them to various online article directories.

Online video promotions: Can promote the OTC drug *via* YouTube channels, Facebook and Instagram, Twitter etc. These online source of media target the customers by labeling the videos appropriately.

Importance of OTC marketing to Doctors: Indian consumers hold physicians in high regard and will have the utmost confidence in any things they recommend. According to a survey, individuals were more likely to acquire OTC medications that were prescribed by doctors than those that weren't. With the development of increasingly scientifically advanced goods, physician communication is becoming more and more important. When the product is promoted through the doctors' personal recommendations, the brand loyalty is increased. Patients are more inclined to select the brand recommended by their physician and stick with it going forward.

The OTC market is seeing intense competition from competitors in the pharmaceutical and fast-moving consumer products industries. Because there are several brands that doctors will prescribe for a given disease, marketers need to amp up their efforts in medical marketing. Where both over-the-counter and prescription medications are effective therapy alternatives, competition is fierce. Most doctors believe that prescribed drugs are more reliable than over-the-counter ones. Therefore, by frequent and thorough dialogue with healthcare experts, marketers should establish a case for its non-prescription products.

Ways to market OTC drugs to Doctors

1. **Offer product samples:** Product samples are an effective way to persuade doctors to suggest a specific OTC brand.
2. **OTC as supporting need**: In a larger treatment plan, doctors may add over-the-counter medication to the prescription to treat related issues. Marketers could explain how their product addresses the majority of symptoms that patients experience.
4. **Be innovative**: Marketers must use a creative strategy because there isn't much area for making scientific proof the focal point of marketing for the majority of OTC items. OTC marketers may benefit from creative campaigns that help customers connect their company with the issues it solves.
5. **Understand mind-sets:** Pharmaceutical corporations assist doctors with selecting and recommending certain OTC medications by carrying out studies on mindset. These research will be useful in identifying prevalent misconceptions and knowledge gaps regarding the items.
6. **Leverage digital media:** The communication between pharma and doctors may be quickly closed by online media that has the confidence of medical professionals. Marketers can use unbiased internet forums where doctors may voice any concerns they may have about OTC medications and get new insights from other experts. Such an internet connection can be beneficial for developing a relationship with physicians and learning the most recent insights about their mindset.

Answer the Following Questions

Q1. Define the term advertising in marketing. Enlist its objectives.

Q2. Enumerate various channels used in advertising.

Q3. What do you understand by sales promotion? Enlist its methods.

Q4. Define Promotional mix. Enumerate the main factors influencing promotion mix.

Q5. What are the most important methods of promotional budgeting in pharmaceutical company?

Q6. Discuss the various personal selling.

Q7. Summarize the steps involved in personal selling.

Q8. Define the terms pre-approach and approach in personal selling.

Q9. What are the obligations and duties of salespeople in the pharmaceutical industry?

Q10. What goals does sales promotion seek to achieve?

Q11. Enlist some promotional tools a medical representative may use in the doctor's chamber.

Q12. What procedures must be followed for the creation and implementation of advertising?

Q13. Write the importance of advertising?

Q14. Describe the use of journal in pharmaceutical marketing?

Q15. What is the importance of retailing in Pharma Industry?

Q16. Explain the role of public relation in pharma,

Q17. Discuss the various ways to market OTC drugs to doctors.

Q18. Give an account on "On line promotional techniques" for OTC products.

Q19. Explain the role and key skills of retail pharmacist.

Q20. Write note on sampling and its impact on product marketing.

MCQs

1. Describe advertising.

   A. Publicity B. Sales promotion

   C. Paid information D. all of the above

2. Which of the following forms the basis of any marketing or advertising effort.

   A. Research,

   B. target segmentation,

   C. a creative brief,

   D. media preparation are all examples of the above.

3. The following terms are used to describe the compensation given to a salesman, agency, etc. as a percentage of their sales:

   A. Replication B. Implication

   C. Commission D. Expansion

4. What word is the root of the word retail?

   A. Latin B. French

   C. English D. German

5. A retailer is a person who sells products in either large or small amounts.
   A. Latin
   B. french
   C. Any of these but
   D. None of these
6. Sales promotions are primarily used to increase sales levels in the:
   A. limited time
   B. Long-term
   C. Medium-term
   D. None
7. Which of these is not a sales promotion technique?
   A. Coupon
   B. Bonus packs
   C. Loyalty card
   D. Questionnaire
8. The best coupon redemption rate achieved is usually in the range:
   A. 6 - 9 %
   B. 3 - 5 %
   C. 10 - 12 %
   D. None
9. The main reason why organizations use exhibitions is to
   A. Have competitive presence
   B. Create publicity opportunities
   C. Make sales
   D. Develop relationships
10. What does OTC stand for?
    A. Other tablets and capsules
    B. Occasional therapeutic care
    C. Over-the-counter medicine
    D. Optional therapies and cure
11. What may happen if you don't follow the instructions on an OTC label?
    A. It could have issues that need for medical attention.
    B. It might also have negative side effects.
    C. It might result in long-term harm
    D. The entire list
12. OTC medicines should be stored:
    A. In a refrigerator only
    B. In a dark closet
    C. Out of reach of children
    D. On the kitchen table
13. Public relation plan can be expressed in how many steps
    A. 4
    B. 2
    C. 3
    D. 5

14. Which of the following promotional budget strategies incorrectly sees sales as the cause rather than the effect of promotion?
    A. An inexpensive way
    B. The percentage-of-sales approach
    C. Method of competitive parity
    D. Goal and task-based approach
15. The promotional mix is the combination of strategies chosen to inform customers about the product or service.
    A. Product
    B. Marketing
    C. Budgets for advertising
    D. Advertising techniques
16. Which of these is not a means of promoting sales?
    A. Free gift
    B. Discount
    C. Budgets
    D. Coupons
17. Among the following which one shows determinants of promotional mix?
    A. Use of product
    B. Type of product
    C. Complexity of product
    D. All
18. How many different approaches are used to establish the budgets for promotional spending?
    A. 4
    B. 5
    C. 6
    D. 8
19. Zero Method is used for
    A. Promotional budgeting
    B. Promotional Training
    C. Promotional Plan
    D. Promotional selling
20. SVL stands for
    A. Standard Visiting list
    B. Standard Versatile List'
    C. Special Visiting List'
    D. Simple Visiting List'
21. A "brand audit" or "prescription audit" conducted at
    A. Chemist level
    B. Doctors level'
    C. Customer level
    D. Wholesaler level
22. POP full form is given by
    A. Point of purchase
    B. Part of purchase
    C. Point of promotion
    D. Point of publicity
23. Traders' (wholesalers and merchants) gifts are referred to as
    A. Deal loaders
    B. Trade deal
    C. Push money
    D. Trade shows

24. Electronic and neon signs are used for
    A. Advertisement B. Trade deal
    C. Market research D. All
25. CIMS can be expressed as
    A. Current Medical Specialties Index
    B. Standardized list of medical specializations
    C. The most recent medical specialty index
    D. The most recent medical store index
26. Subhiksha, a South India based retailer follow which type of format.
    A. Discount store B. Discount merchant
    C. Convenience store D. Superstores
27. HCPs can be unlocked as
    A. Heath care professionals B. Heath care professors
    C. Heath consultant professionals D. Heath care people
28. Bring your outdated colour television so that you may get a Samsung LCD TV for Rs. 15,000 and take it home.
    A. Exchange Schemes B. Finance Schemes
    C. Price promotion D. Combination Promotion
29. What kind of deal is it if you receive Rs. 30 off and a multifunctional jar for free when you buy a one-liter bottle of Vatika herbal shampoo?
    A. Multiple promotion B. Finance schemes
    C. Price promotion D. Combination promotion
30. Which strategy involves paying traders more to encourage them to reach predetermined targets?
    A. Push money B. Trade deals
    C. Trade shows D. POP

Answer key to MCQs

1. D	2. C	3. C	4. A	5. B	6. A	7. D	8. B	9. D	10. C
11. D	12. C	13. A	14. B	15. D	16. C	17. D	18. A	19. A	20. A
21. A	22. A	23. A	24. A	25. A	26. A	27. A	28. A	29. A	30. A

CHAPTER 4

Pharmaceutical Marketing Channels

4.1 Introduction to Pharmaceutical Marketing Channels

The individuals, groups, and actions required to transfer ownership of the commodities from the point of production to the point of consumption form the framework for the marketing channel. This is a way out by which the product reaches to the consumer, thus, whole process can be also called as *distribution channel.*

4.1.1 Designing Channel

Decisions involving in the development of new marketing channels where no one had existed before, or the modification of existing channels is said to be designing of a channel or *designing channel.* Producers and manufacturers, wholesalers, and retailers all face channel design decisions. Manufacturers and producers frequently "look down" the channel. While wholesaler intermediates deal with channel design from both angles, retailers "look up" the channel. Channel design refers to the choices made while creating new marketing channels or altering those that already exist. The channel design decision can be framed into six steps mentioned in **Figure 4.1**.

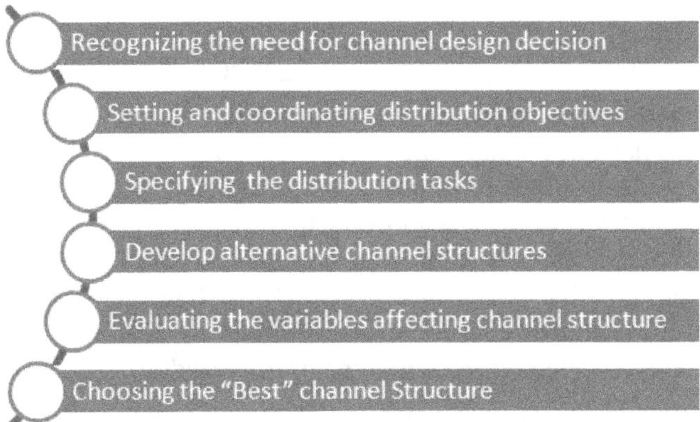

Figure 4.1 Steps involved in the channel design

1. **Recognizing the need for a channel design decision:** the development of a new product or product line, the targeting of an existing product at a new target market, or other circumstances indicating the need for a channel design decision. This will also make a major change in the component of the marketing mix. The channel design is helpful in establishing a new firm. It also adapts for changing intermediary policies and dealing with changes in availability of particular kinds of intermediaries. The channel design will open up new geographic marketing areas and meet the challenge of conflict or other behavioural problems. Finally, this will be helpful in reviewing and evaluating.
2. **Setting and coordinating distribution objectives**: In order to setup the distribution objectives with other marketing objectives and strategies, the channel manager needs to perform three tasks:
   (a) **Familiar with objectives and strategies:** The firm's goals and tactics may have an influence on the distribution goals. In reality, the same person or people that establish the goals for the other elements of the marketing mix frequently also do so for the distribution channel.
   (b) **Setting explicit distribution objectives:** In essence, distribution objectives are declarations that explain how distribution helps a company achieve its overall marketing goals.
   (c) **Checking for conformity:** A conformity check verifies that the distribution objectives do not conflict with the other areas of the marketing mix.
3. **Specify distribution tasks**: It is appropriate to decide on the precise distribution duties (functions) to be carried out in the system once the objectives have been established. The duties must be described in detail by the channel manager, who must also explain how they will alter based on the circumstances. For instance, a manufacturer may list the following requirements as essential to successfully reach the target market, such as offering delivery within 48 hours of an order being placed.
4. **Develop alternative channel structures:** To complete their distribution duties, the channel manager should take into account alternate methods of distributing distribution objectives. The channel manager may frequently pick many channel structures in order to successfully and efficiently reach the target audiences. The allocation options (potential channel structures), whether they involve one or more channel structures, should be assessed in terms of the following criteria: ***as number of levels in the channel, intensity at the various levels and type of intermediaries at each level.***
   (a) **Number of levels:** Channels can have two to numerous layers (five is typical). Direct channels include the two-level producer-to-consumer route. Due to history, all businesses in one industry may have the same number of levels, however in certain other industries it may vary.

(b) **Intensity at each level:** The channel manager must select the precise amount of channel components utilised at each level after deciding on the number of levels. How many stores should be part of the distribution network in a given market? how many distributors? The choice of intensity is particularly crucial since it plays a significant role in the company's entire marketing strategy. As an illustration, businesses like Starbucks and Hershey's have had significant success thanks to their rigorous distribution strategies.

(c) **Types of intermediaries:** Different sorts of intermediates may be found in any channel system. The goal is to find several potential alternative channel architectures and assess them using a variety of characteristics, including business-related aspects, environmental trends, a reseller's reputation, and a reseller's expertise.

5. **Evaluating the variables affecting channel structure:** Just as crucial as evaluating the other marketing responsibilities is evaluating the performance level of the channel members. The marketing mix is obviously very interrelated, and the failure of one element might result in the collapse of the entire strategy. The channel manager should then assess various factors to see how they could affect different channel architectures. The following variables are used to assess channel design:

   A. **Market variables:** The most important factors to take into account when developing a marketing channel are market variables. Subcategories of market factors have a significant role in affecting channel structure. They are ***market geography, market size, market density, and market behavior.***

   (i) **Market geography:** refers to the markets' geographic size, physical position, and separation from the producer and maker. "The greater the distance between the manufacturer and its markets, the higher the possibility that the use of mediators will be less expensive than direct distribution," is a well-known heuristic (rule of thumb) for connecting market geography to channel design.

   (ii) **Market size:** A market's size is determined by how many consumers or businesses make up the market. The market size increases with the number of unique clients. Due to the high transaction fees of supplying numerous individual clients, the employment of mediators is much more likely to be essential in big markets. On the other hand, a company is more likely to be capable of avoiding using middlemen if the market is small.

   (iii) **Market density:** The market density is determined by the number of purchasing units per unit of land area. Distribution is typically more complex and costlier the less dense the market. "The less intense the market, the more probable it is that mediators will be utilised," is a heuristic for determining market depth and channel structure. On the other hand, the possibility of removing intermediaries increases with market density.

(iv) **Market Behavior:** It alludes to the following consumer purchasing patterns that How do customers buy? When, where, and from whom clients make purchases are all important information. Each of these purchasing habits may have a considerable impact on the channel structure.

B. **Product variables:** Product variables such as ***bulk and weight, perishability, unit value, degree of standardization (custom-made versus standardized), technical versus non-technical, and newness affect alternative channel structures***.

(i) **Bulk and weight:** Large, heavy things have high handling and transportation costs relative to their value. Therefore, a producer should try to lower these costs by only sending in large numbers to a select few locations. As a result, the channel topology from producer to user should normally be as short as is practical.

(ii) **Perishability:** Products that degrade quickly physically and those that go out of style quickly need to move quickly from manufacturing to consumption.

(iii) **Unit value:** The channel should really be longer the lower the product's unit value. This is due to the little margin for distribution expenses left by low unit value. Direct distribution is possible when the value is high in relation to the product's size and weight since the transportation and handling expenses are low in relation to the product's worth.

(iv) **Degree of standardization:** While more standardised items give the chance to expand the channel, products should move from manufacturer to customer.

(v) **Technical versus Non-technical:** A highly specialized product will often be supplied through a direct route in the industrial sector. This is because the producer may require sales and service personnel who can explain the technical aspects of the product to the customer. For the same reasons, relatively technical goods are often sold through limited channels in the consumer market.

(vi) **Newness:** In order to create demand for new items, both commercial and consumer, vigorous and broad advertising is needed. It is typically harder to obtain this type of promotional strategies from all channel members more longer the distribution channel is. Because a well chosen set of mediators is more likely to supply active advertising, a shorter route is typically seen as a benefit for new items.

C. Company variables: The most important company variables affecting channel design are its ***size, financial capacity, managerial expertise, and objectives and strategies***.

(i) **Size:** In general, a firm's size positively affects the variety of alternatives for various channel topologies. Larger businesses have access to more alternatives than smaller businesses.

(ii) **Financial capacity**: Generally speaking, a company's reliance on intermediaries decreases with increased capital available to it.

(iii) **Managerial expertise:** Channel design must unavoidably involve the services of intermediaries with this knowledge for businesses without the administrative capabilities required to carry out distribution duties. It could be possible to adjust the structure over time to lessen the need on intermediaries as the firm's management develops expertise.

(iv) **Objectives and strategies:** The employment of intermediaries may be constrained by the marketing and overarching goals and strategies of the company, such as the need to exert strong control over the product. The sorts of channel architectures that are accessible to those organisations using such techniques will be constrained by their emphasis on aggressive advertising and quick response to changing markets.

D. Intermediary variables: The primary intermediary factors that influence channel structure are **availability, costs, services offered.**

(i) **Availability**: Channel structure will be influenced by the availability of sufficient intermediaries (both in terms of quantity and expertise).

(ii) **Cost:** Choosing a distribution channels always takes the cost of utilising intermediaries into account. The configuration is likely to reduce the use of mediators if the expense of utilising them is too high compared to the services provided.

(iii) **Services**: This entails assessing the services provided by various intermediaries to determine which ones can handle them most efficiently and affordably.

E. Environmental variables: There are various environmental variables as economic, socio-cultural, competitive, technological and legal environmental forces which have a significant impact on channel structure.

F. Behavioral variables: The channel manager should review the behavioral variables. The channel manager makes sure there is a solid foundation on which to persuade the channel members.

6. **Choosing the "Best" channel structure:** The channel manager should select the "Best" channel structure for channel design in order to achieve the desired level of effectiveness in carrying out the distribution tasks at the lowest cost.

4.1.2 Channel Members

The people or individuals who participate in the market's purchasing and selling of products and services are known as channel members. There are three types of **channel members** and named as ***Agents (Buying/selling agent and commission agent), Wholesalers, and Retailers***.

There are various channels of marketing and selection of an appropriate channel depends on the requirements of manufactures, consumers and nature of products. Distribution is the process of transferring products from place of production into the hands of consumers at right time.

4.1.2.1 Number of Channel Levels

The producer is the first stop in the distribution chain, and the client is the last. A channel level is each intermediary layer that contributes to getting the product nearer to the ultimate layer. There are various channel levels:

Channel 1: Manufacturer – Consumer (Direct marketing channel).

Channel 2: Manufacturer – Retailer-Consumer.

Channel 3: Manufacturer - Wholesaler– Retailer- Consumer.

Channel 4: Manufacturer - Distributor- Wholesalers–Retailer -Consumer.

4.1.2.2 Classification of Channel Members

The channel members are named by various names as distribution channel members, middlemen, thus they have classified as:

A. **Functional middle man**: These are the intermediaries who perform task of transferring goods from producer to consumers about acquiring ownership rights and includes many as:

   (i) **Brokers** (Buying or Selling agents): They are the agents who negotiate purchase and sale of goods on behalf of others. They are called selling agents if they are engaged by seller and buying agents when engaged by buyers to negotiate the purchase of goods on their behalf. They get a certain percentage of money on the business transacted by them.

   (ii) **Commission agents**: They sell goods on behalf of seller and take the goods from the seller for a certain time till the negotiation for selling is in progress and makes arrangements for selling the goods to consumers.

B. Merchant middle man: They acquire and transfer the products in their own name:

(i) **Wholesalers**: These are the middleman who act as intermediate between manufacturer and retailer. They buy products in bulk from manufacturer and sell it to retailer (chemists). There are various types of ***Wholesalers*** and commonly named as**:**

(a) **Manufacturer wholesalers**: They produce as well as act as stockiest for their products and transfer their product to retailers.

(b) **Retailer wholesalers**: They purchase goods in bulk from manufacturer and also sell them to consumers through their own shop.

(c) **Distributors (Actual wholesalers)**. They concentrate only on buying products in bulk from manufactures and sell it to retailers.

Function of wholesalers:

- **Assembling**: Wholesaler buys large variety of goods from manufactures and sells them in small quantities to retailers.
- **Distribution**: They distribute goods to retailers who are generally widely scattered.
- **Warehousing**: The goods purchased from manufacturer are kept in stock in warehouses of wholesaler till these are distributed to retailers.
- **Transportation**: Wholesaler has to move the goods from place of production to his own warehouse.
- **Financing**: Wholesaler provides credit facilities to the retailers and thus finances the retail trade.
- **Risk bearing**: Buying in bulk and storage carries a no of risks such as changes in demand, spoilage of products or loss due to theft in warehouses. All such risks are owned by wholesaler.
- **Packaging**: Some wholesaler also performs the function of packaging of products into smaller lots for retailer.
- **Grading**: Division of the products according to their grades has been done by wholesalers.
- **Market research**: Done by producer/wholesaler. Wholesaler is close to retailer and retailer knows the needs of customers. Thus wholesaler gets information about needs of customers from the retailers and can advise the manufacturers to design the products accordingly.

(ii) **Retailers**: Retailer is a middleman between wholesaler and consumer. Retailers dealing in pharmaceutical trade are chemist or pharmacist. Retailers buy the product from wholesalers and sell it to consumer. A retailer is someone who makes it their business to offer consumers a wide range of items that are assembled on-site to meet the demands of end users.

Types of retailers:

(i) **Itinerant retailers**: These type of retailers don't move from fixed business premises but move from place to place as hawkers and market traders.

(a) **Hawkers**: These retailers move from door to door in residential localities to sell their goods.

(b) **Market traders:** They keep on moving from place to place to sell their goods at weekly/monthly markets or at annual trade fair.

(ii) **Fixed shop Retailers**: They are either small scale fixed retailers or large scale fixed retailers as:

(a) **Small scale fixed retailer:**

Street stalls: These stalls are located at a street crossing or in the main street. Products are sold in a small platform on a very scale.

Fixed shops: These are the fixed shops stocked with a variety of products of various brands but in small scale

(b) **Large scale fixed retailers:**

Retail departmental stores: It is a sizable retail organisation made up of a number of departments, each of which deals in a distinct line of products. All of these departments are housed under one roof and are managed by the same person, with the benefits and drawbacks listed in Table 4.1.

Table 4.1 Merits and demerits of Retail departmental stores

S.No	Merits	Demerits
1.	Provides convenience to customer.	High cost of running a department store.
2.	Customer get products at one place.	Sales are controlled by employees.
3.	Large variety of goods offers a good choice to customer.	Stores can't establish personnel contact with customers.
4.	Bulk of product range give high profits.	Due to working independently may not provide co-ordination with other departments.
5.	Free home delivery to customers.	Price of products charged is usually high.
6.	Afford competent salesman to attend the customer.	
7.	Provide various facilities to customer like telephone and recreation facility for customer satisfaction	

Multiple shops/ Chain stores: These are the groups of shops in the same branch of retail trade. A single business firm opens a number of shops that are situated at different localities in the city or in different parts of country. All branches deal in similar types of products. Each multiple shop system has a central office, from which all of its branches get decisions. Every branch charges the same price, which is established at the corporate level for all goods. Items are supplied to all branches, such as V-Mart, Easy Day, etc., from the main branch. They have a few advantages and disadvantages, which are listed in Table 4.2 below.

Table 4.2 Merits and Demerits for Multiple shops/Chain stores

S.No	Merits	Demerits
1.	Supplies are made by main branch to other branches.	Staff at multiple shops has little freedom to make its own decision that may sometimes adversely affect the sales.
2.	Bulk buying is performed, so it's a profitable business	Centre has little control on sales volume.
3.	Located centrally in the city, make convenient purchase to customer.	There is no provision for any personal special facilities to customers.
4.	Fixed price and standard quality of products.	
5.	Easy identification by uniformity in window display.	
6.	Payment through cash or credit cards.	
7.	Uniform policies are followed in all branches so convenient to customers	

Mail order business: It is a type of retail trade where all business activities take place by post. The manufacturer sells their products directly to customers without involving middleman (Wholesaler and retailer). Manufacturer approaches to the expected customers by sending the price catalogue and product circular by post or by means of advertisement in newspapers/ magazines. If customer's places order for supply of goods manufactures sends the product by V.P.P (Value payable post). Under VPP the goods are delivered to the address on payment of price to the post office which sends the money to manufacturer. Mail order business have few merits and demerits, which have been mentioned below in Table 4.3.

Table 4.3 Merits and Demerits for Mail Order Business

S.No	Merits	Demerits
1.	System doesn't need a shop/showroom to start the business.	Buyers can't examine the product before its purchase.
2.	Running expenses are low, as system doesn't require a large no of employees.	Risk of damage of products is there during transportation.

Table 4.3 *Contd....*

S.No	Merits	Demerits
3.	Useful for those customers who live in remote areas	Lack of personal contacts between manufacturer and consumer.
4.	No middle man is required so these expenses are saved.	There are chances of cheating by dishonest producers.
5.	There is no danger of loss due to debts.	Wastage of time and money in correspondence.
6.	Price of products is either received in advance or collected from post office.	

Functions of retailers:

- **Keeping the stock**: A retailer keeps a ready stock of all the goods. So that he is able to meet the demands of his customers at all times.
- **Transport**: Retailers make their own arrangement for transportation of goods from wholesaler's warehouse to their retail stores.
- **Risk bearing**: retailers have to maintain a reasonable stock of goods so that the consumers are satisfied. As far as the items are not sold, they are responsible for any loss due to fire, theft, or degradation.
- **Financing**: Retailers assist consumers financially by offering them things on credit.
- **Market research**: They are in close contact with customers. They know the taste, likes and dislikes of customers. They pass on such information to wholesaler for onward transmission to the concerned producers. Thus they help in market research.
- **Sales promotion**: They can provide special schemes to the customers, thus they help in promoting sales. They can also advice the customer about benefits of the products and remove their doubts of consumers. Thus they promote sales.

4.1.2.3 Strategies for Best Marketing Channel

Develop a marketing strategy: Strategy is everything in marketing and among hundreds of marketing channels (PR, influencers, events, digital, print, newsletters, etc.), Selecting those that can be done with less money is more crucial. Starting with a few channels where customers are most likely to perceive messages and act, a strong marketing strategy places strategic investments (like, share, buy, comment, sign up, review). By engaging the consumers at every stage of their purchasing journey, all the way from awareness to purchase intent, to taking action on the purchase itself is a marketing strategy.

Consumer-first mindset: The mind-set of a consumer is visualized by the company for developing marketing channel which can drive them to purchase their product and can recognize where the consumer spends their time for online purchases. This can also help to drive strategy in the selection of marketing channels.

Compare yourself to others: What do you hope to accomplish? Do you desire folks to purchase? Would you like people to know? Different marketing channels are more effective at reaching various marketing goals than others. Different digital tools, like social media advertisements enable to position the message in a way to achieve a specific goal, which can be purchased, or simply to grow the customers.

4.1.2.4 Function of Channel Members

Main functions of channel members in channel distribution consists:

Communication: Marketing intermediaries promotes the company's products. Here, the channel member provides the information regarding the product and pushes it to customer.

Title transforming: Marketing intermediates buy the products from the business and transfer the ownership of the products to the following middleman or client.

Relationship management: The marketing intermediates work to comprehend, match, and fulfil the demands of the consumer.

Risk taking: When a distributor or channel member is buying a product, and it does not sell out, or the distributor suffers bad debts or any troublesome thing happens, then the risks will be taken by channel member.

Financing: The majority of businesses work with credit limits or advance payments. Therefore, it is not required that the channel participant receives payments from clients within that time frame. A business could accept a down payment for product X, but that product might end up being sold after 40 days. Therefore, the channel member bore the economic strain of the product for the first 40 days. Members of the channel should be prepared for such financing.

Physical distribution of goods: There are several ways to contact the final client when any channel is the driving force behind a business. Delivering the goods to a channel member is the company's responsibility. However, it is the duty of channel participants to make sure that the products are delivered to the final buyer as soon as possible and in the best possible condition.

Negotiations: Members of the channel conduct all discussions with the final consumers; the corporation is not involved. When a channel member purchases a product, the product becomes their property, and the channel member is also responsible for the sale of that product.

Inventory management: The distributor or reseller must match the inventory that the market requires and that is on hand. The dealer's inventory and finances will be hampered if someone places a pointless order for material that is not now in demand.

Contacts: The work of multichannel dealers entails keeping in touch with current clients as well as developing relationships with future clients and sharing them with the business. Good businesses also provide their dealers access to a CRM (customer relationship management) system, which they may utilise to better retain customers.

Promotions: In addition to being conducted at the company or brand level, promotions are also done at the channel level. A distributor or distributor will employ promotion and advertising to draw customers to them whenever he wants to increase brand recognition and inform customers about the brand's point of purchase. In reality, channel members have a built-in responsibility to boost local sales.

Information: A role of a channel member is to gather data about future clients, the competitors, and environmental conditions. The involvement in making marketing strategies for the company is incomplete without information from the channel member and the company cannot move in the right direction.

4.1.2.5 Decision in Setting up a Channel

Marketers will choose a channel that is best for their company when making their decision.

Understanding the customer profile: Everybody has a different set of buying habits. People who are time-constrained prefer to make purchases online, whereas those who are time-rich prefer to go shopping. While some of them like to have a range of things, others choose specialty or unique goods. Therefore, marketers need to know who their target audience is. What do they buy?

Objectives to develop a channel: It is determined by the ***reach, profitability, differentiation.***

(i) **Reach**: The company wants to have the products available in the majority of retail establishments. It will use an active distribution strategy.

(ii) **Profitability**: The company wants to increase profitability while lowering channel costs.

(iii) **Differentiation**: Different product positionings are used by the company. While the majority of the industry participants use the traditional approach, the firm uses a new channel model. Example: All computer companies used the dealer/retailer route to market their goods, but only Dell began doing it online.

Identify type of channel members: Following the establishment of objectives based on company policies, an analysis of the most appropriate channel type will be conducted. Sometimes intermediaries are referred to as retailers, agents, and resellers.

Intensity of distribution: Intensity of distribution means how many middlemen will be used at the wholesale and retail levels in a particular territory. The company may choose from any of the following approaches depending on the number of intermediaries needed.

(i) **Intensive distribution**: a business approach that involves stocking products at several venues. The goal is to place the products close to the buyer. For instance, Parle-G glucose biscuits are sold in virtually all retail establishments in both urban and rural areas.

(ii) **Selective distribution**: a business approach where products are stocked in a small number of retail locations. Example: Only a few retail locations sell televisions.

(iii) **Exclusive distribution**: This is a channel format type when marketers gives only limited numbers of dealers the exclusive right to distribute its products in their territory. Example: Tanishq a jewellery outlet, BMW etc.

Responsibilities of channel members: Company should define the territory in which channel members should operate, at what price it should sell, and provide services.

Criteria to evaluate the channel members: A company could provide a variety of channel options. It wants to select whichever choice best satisfies its goals. The SCPCA approach may be used to assess channels at the design stage.

(i) **Sales (S):** the capacity of each network member to produce sales for the business over a specific time period.

(ii) **Costs (C):** How much does each alternate channel cost? Which of the possibilities yields the best result?

(iii) **Profitability (P):** The profitability of the company's available channel options will be compared. The company that is more profitable will be chosen.

(iv) **Control (C):** Every business wants more authority over the channel participants. Depending on how much power each channel member wants, different channels can be compared? and how much authority the business is prepared to give up.

(v) **Adaptability (A):** Companies are under constant pressure to review their procedures and supply chain due to competition. The channel choices must to be adaptable enough to satisfy shifting needs. Whichever options satisfy these goals should be chosen.

4.1.3 Selecting the Appropriate Channel

Distribution channels play a significant role in the "marketing mix" of any company. Because various components of the marketing mix, such as price and promotion, are closely related to and dependent on the distribution channels, choosing the right distribution channel is crucial.

A large number of middlemen are available through which a product can be distributed. Each middleman involves some cost which enters into the price of the product that the ultimate consumer has to bear. If a wrong choice of distribution channel is made, it will lead to increase in the distribution cost which will lead to either lowering down of profits or increasing the cost of the product to the customer.

However, both the alternatives are not good for the long run survival and growth of the business firm. Therefore, selecting the appropriate distribution channel is crucial. The manufacturer can only sell his items through the proper channels at the correct time and in preset quantities. Selecting the best distribution channels is not an easy task.

Decision for selection of appropriate channel:

(a) **Market considerations:** The nature of the market is the key factor influencing the choice of channels of distribution. This is determined by the following:

   (i) **Consumer or Industrial market:** If the product is intended for industrial market or industrial users, the channel of distribution will be a short one. Since, industrial users purchase in large quantities, they can purchase directly from the producers or manufacturers, i.e., there is no need of retailers. The manufacturer can establish contacts with the industrial users by sending his agents. In case the product meant for the consumers, retailers may have to be included in the channels of distribution.

   (ii) **Number of potential customers:** A large potential market is likely to put weight in favour of the use of middlemen. If the number of customers is relatively small, Due to the ease of handling customers, the producer may be capable of selling effectively through his own sales team.

   (iii) **Size of order:** Where consumers submit orders in large amounts, as in the case of industrial items, direct selling is practical and cost-effective. But middlemen are utilised to distribute such things in cases where the commodity is sold in modest amounts. The same producer may choose various distribution methods for his goods. For instance, a producer of food items may offer directly to large retail establishments and employ wholesalers to reach small shops.

   (iv) **Buying habits of customers:** The customer's purchasing habits, such as the amount of time he is willing to invest, his desire for credit, his preference for personalised service, and his preference for one-stop shopping, have a big impact on the distribution channels chosen. Example, candles are purchased in ones and twos, and rarely in packets. This calls for need for retail stalls.

(v) **Geographical concentration of market:** Where prospective customers are located in a particular geographical region, direct selling is more feasible than it would be, if the market were spread over the whole country. Use of wholesalers and retailers may become essential to sell the product to the widely dispersed consumers or industrial users.

(b) **Product Considerations:** The type and nature of the product influences the number of type of middlemen to be chosen for distributing the product. The factors with regard to the product are as follows:

(i) **Unit Value:** If the unit value of the product is lower and the turnover is higher, the channels of distribution will be longer. For instance, products like cosmetics, stationery and small accessory equipment are distributed through agents, wholesalers and retailers. Products of high value like jewellery and industrial machines are sold directly to the users.

(ii) **Product line:** A manufacturer manufacturing several products in the same line will sell directly or through retailers since it is economical. But a manufacturer with only one item may have to use wholesalers and retailers to sell his product.

(iii) **Standardised product:** Standardised products can be distributed through longer channels because their brand names are very popular. But custom made and un-standardised products can more easily be sold directly by the producers to the user.

(iv) **Technical nature:** Industrial products which are highly technical, as air conditioners and super computers, are often distributed directly to the industrial users. The manufacturer of such a product can appoint sales engineers who can explain the product to the potential customers and provide presale and after sale service to them.

(v) But consumer products of technical nature are generally sold direct and through retailers. This is because of the fact that consumer products are sold in large number and it is not feasible for the manufacturer to provide after-sale service to the consumers as in case of television sets and refrigerators.

(vi) **Bulk and Weight:** Bulky and heavy goods are distributed directly to users in order to minimise the physical handling of the product because transportation of such a product involves huge cost.

(vii) **Perishability:** The channels of distribution are short in case of products subject to physical decay or fashion change. The producers of perishable products generally sell directly to the consumers or sell through the middlemen who have the special storage facilities.

Manufacturers of non-perishable commodities have a wider choice in the channel selection.

(c) **Company considerations:** The choice of distribution channels is significantly influenced by the kind and scale of the commercial entity. In this sense, the following criteria are crucial:

(i) **Volume of production:** A big manufacturer may find it profitable to sell directly to customers through the sales force. If it is manufactured in wide range of products, it may sell out the products by opening retail outlets in different parts of the country. But a small manufacturer with only a small number of items cannot afford to sell directly because of small scale operations.

(ii) If a big manufacturer has manufacturing units scattered in different parts of the country, it is more economical to engage wholesalers and retailers to sell out the products.

(iii) **Financial resources:** A financially strong company can distribute products by employing its own sales force and opening retail outlets. But a financially weak company which cannot invest money in distribution will have to use middlemen to sell its output.

(iv) **Experience and competence of management:** If a company's management is having sufficient experience and know-how to market its products, preference should be given to distribute products directly. A company lacking this ability and experience will often rely heavily on middlemen to do the marketing job.

(v) **Services provided by the channels:** The services offered by the manufacturers have an impact on the distribution channel choice as well. Only if the marketing department invests enough in promotion will a manufacturer be able to identify reputable merchants. Some makers of technological goods promise to offer post-sale support. Such a manufacturer can also obtain reputable merchants to market its goods.

(vi) **Desire for control of channels:** A manufacturer who wants to control the distribution of the product will select a short channel of distribution. The distribution cost is higher if the marketing department can give aggressive promotion to the product.

(d) **Middlemen considerations:** The following are some middlemen-related variables that affect the channel choice:

(i) **Availability of desired middlemen or intermediaries:** If a manufacturer runs their business as they see fit, they will rely on middlemen. The marketer may not trust product to a middleman who is carrying competitive products. In such a case, it may prefer to open branches to sell products directly.

(ii) **Financial ability:** A large manufacturer will generally select those middlemen who are financially strong, can provide credit facilities to the customers, and pay their bills to the manufacturer regularly and promptly.

(iii) **Attitude of middlemen:** Sometimes, middlemen are not prepared to carry a manufacturer's product because of the non-acceptability of marketing policies. For instance, a retailer may want sole selling agency for a particular region or a guarantee against price reduction for handling the products of the manufacturer.

(iv) **Sales potential:** A manufacturer will generally select a channel offering the greatest potential sales volume over the long-run, though it is very difficult to assess which channel will generate the largest sales volume.

(v) **Cost:** The manufacturers also consider the cost of selling through alternative channels. It does not mean that a middleman charging high cost would be excluded from consideration. A manufacturer may select even a higher cost charging middleman who provides many services to the customers which are not provided by other middlemen. This would provide added value to the customer.

(vi) **Competition and legal constraints:** Many times, the manufacturers are compelled to use the same channels of distribution which are being used by the competitors. Government regulations also affect the choice of middlemen. For instance, a pharmaceutical company can market its product through licensed chemists only.

4.1.4 Conflict in Channels

Any disagreement, conflict, or struggle between two or more sales channels in which one partner's activities have an effect on the other partner's firm, sales, prosperity, market share, or accomplishment of a comparable objective is referred to as a "channel conflict.".

To get its products to customers, a business creates a network of dealers and distributors. However, this network will always experience friction since it is manually operated. This friction manifests as channel conflict, which develops for a number of reasons. Price differential and territorial encroachment are two possible causes. The numerous sorts of channel conflicts must be understood by a firm that sells its goods through channel marketing in order to handle channel conflicts.

4.1.4.1 Types of Channel Conflict

According to its flow and the parties engaged, the channel dispute may be divided into the four categories listed below:

Conflicting horizontal channels: When businesses are located at the same top of the channel, there is horizontal conflict. Horizontal channel conflicts are among the most frequent types that happen. A horizontal channel conflict occurs when two actors in the distribution channel are competing at the same level. Therefore, a disagreement between two distributors or two retailers is referred to as a horizontal channel conflict. A horizontal conflict could arise, for instance, when two franchisees build two restaurants next to one another on the same street or when one company in a supply chain provides cheaper pricing than the other companies in the supply chain and so draws in more clients. Conflict between channel partners of same level, i.e. disputes between two or even more stockists or retailers from different regions about price or manufacturer prejudices, is referred to as horizontal level conflict.

Vertical channel conflicts: Vertical channel conflict is another sort of conflict that can occur in channel management. Vertical channel conflict occurs at various levels of the distribution platform, as opposed to horizontal channel conflict, which occurs between participants at the same level. A common dispute might arise between the distributor and the store or between the distributor and the buyer/ carrying and forwarding agents (C&F) and the company. For instance, channel conflict between dealers and retailers or wholesalers and retailers. Example: Suppose, an Ice cream company motivates its dealers, by providing FREEZERS at discounted price along with the ice cream to their retailers.

Because horizontal conflicts happened at a lower level than vertical channel conflicts, addressing vertical channel conflict is far more challenging for businesses. Vertical channel disputes, however, may also include the distributors or the manufacturer. Management of vertical channel disputes is therefore crucial for the organisation.

Inter-type channel conflict: Conflicts of this nature sometimes occur in scrambled retailing, where major retailers make a special effort to enter a line of products that differs from their typical product range in order to compete with the tiny and concentrated stores.

Multi-channel level conflict: When a producer sells its goods through numerous channels, it may have a conflict at the multi-channel level where channel partners interested in one distribution channel run into problems with another channel. Such conflict arises when a company possess multiple channels may have a retail store, and e-store, a wholesale and hypermarket distribution centre. Each component is a channel whereby a customer can receive products or services. Due to bulk selling power e-store and hypermarket like mall give huge discount which diminish the customer loyalty for other channels.

4.1.4.2 Conflict Magnitude

The severity of the disagreement determines whether it is regarded significant or requires the channel leader's as well as the manufacturer's attention. By properly analysing the shift in market share as well as the company's volume of sales in a certain area or region, the severity of the dispute may be ascertained.

4.1.4.2.1 Causes of channel conflict

Following are the key reasons for an organization where they face channel conflict:

(i) **Role ambiguity**: In a multi-channel setup, this unreliable action by an intermediary might disrupt the distribution channels and result in conflict amongst the intermediaries.

(ii) **Incompatible goals**: Conflict in the channel occurs when the producer and the intermediaries have distinct goals and attempt to achieve them in different ways.

(iii) **Marketing or strategic mis-alignment**: Often, two channel partners will market the same manufacturer's goods in divergent ways. This will give consumers two different impressions of the same product, which can lead to contradicting brand perception.

(iv) **Difference in market perception**: The manufacturer and the intermediaries can have differing perspectives on the potential market, which might cause conflict and diminish the intermediaries' willingness to take control of that specific market.

(v) **Change-resistant:** The channel leader's intentions to reorganise the distribution route may not get support from the intermediates. As a result, it could result in discord or lack of collaboration.

(vi) **Inappropriate geographic and demographic distribution**: When the route leader authorises several selling partners but the sales zone has a tiny consumer base, they occasionally lose interest fast due to the low profit and low volume of sales.

(vii) **Goal incompatibility**: Conflict arises when diverse parties in the distribution route have objectives which might or might not be similar.

(viii) **Ambiguous roles**: The channel partners could not understand their function clearly, including what they are expected to perform, which markets to serve, what pricing strategy should be used, etc.

(ix) **Different perceptions:** Conflict may arise because the channel partners' differing viewpoints on the market conditions hurt the company as a whole.

(x) **Lack of communication**: One of the main causes of disagreement between channel partners is due to this. Any modifications that are not promptly conveyed to all partners would impede the marketing process and lead to discrepancy.

(xi) **Manufacturer dominating the intermediaries:** For the producer, the process of distributing goods and services is carried out through intermediaries like wholesalers, distributors, retailers, etc. Conflict develops if the producer instantly changes the pricing, the product, or the marketing strategy.

4.1.4.2.2 Consequences of channel conflict

It's important to comprehend the root causes of these disagreements and how hazardous they might be for a business. There are few outcomes of these conflicts seen:

(i) **Price wars**: Because of channel conflict, where partners compete with one another on price, a consumer may put off making a purchase while looking for the best price.

(ii) **Customer dis-satisfaction**: If there is a channel conflict, distributors or merchants could be too interested in the company's goods and unwilling to help customers, which makes them resentful of the brand.

(iii) **Sales deterioration**: Conflicts can have a negative effect on sales of the items since distributor interest is declining and more customers are switching to rivals' products.

(iv) **Distributors exit**: Manufacturers must maintain their distributors or business partners in order to increase product sales. There is a higher chance that various distributors may abandon the channel if there is a network disagreement.

(v) **Bad public relations:** Because of the company's bad connection with the distributors, disgruntled distributors may advertise the brand and its products negatively.

4.1.4.2.3 Channel conflict management

Although disputes can never be completely settled, they can be managed to have less of a negative impact on a company. Here are several methods for dealing with channel conflicts:

(i) **Diplomacy, arbitration and mediation:** One representative from each side of the issue is sent by the parties to mediate the situation diplomatically. Both parties agree to submit their reasons to the arbitrator during the arbitration and to the outcome. Through his conciliation abilities, the third party becomes involved in the mediation process to resolve the disagreement.

(ii) **Exchanging employees:** Swapping personnel across levels is one of the greatest strategies to avoid channel conflict; for example, two or more people can temporarily go from the manufacturing level to the dealer level and from the wholesale market to the retailer level. As a result, role ambiguities are reduced and everyone is aware of each other's functions.

(iii) **Co-optation:** As a member of both the complaint handling committee or boards of directors, the manufacturer should choose a specialist who has prior experience handling channel disputes in other organisations in order to handle such conflicts. In order to eliminate conflicts via their professional opinions, any executive or subject matter expert from another organisation may be placed on the advisory committee, the board of directors, or complaint handling committees.

(iv) **Dealer councils and trade associations:** The manufacturer creates a dealer forum where the dealers may collectively voice their issues and complaints to the channel leader in order to address horizontal or vertical disputes. They can be included as members of a trade association that protects their interests in order to foster unity among the sales channels or intermediaries. Creating a connection amongst the channel partners is another strategy for resolving the issue inside the channel. Through the intermediates' shared membership, this is possible. Each channel partner acts in unison and as a single unit.

(v) **Subordinate goals:** In order to prevent disputes, the intermediaries must settle on a common objective in terms of increasing market share, survival, maximize profits, high quality, customer happiness, etc.

(vi) **Superior goals:** By identifying the organization's overarching purpose and synchronising it with the distinct aims and goals of the channel partners, channel conflicts may be minimised.

(vii) **Regular communication:** The channel leader should regularly seek feedback from the channel partners in order to understand dynamics and trends. Additionally, issues and conflicts with the distributor may be overcome by speaking to one another often.

(viii) **Legal procedure:** If the problem is significant and the channels leader is unable to handle it, the offended person may bring a lawsuit against the defendant.

(ix) **Fair pricing:** Since the price war is the primary cause of most channel disputes, it is possible to resolve them by ensuring that products are priced competitively across all regions and that channel partners receive a fair margin.

4.1.5 Physical Distribution Management

Controlling the transfer of resources and items from their source to their destinations is a component of physical distribution management (PDM). The actual transfer of commodities from of the producer to consumer is referred to as physical distribution. It is a crucial aspect of marketing efforts and a key element of the marketing mix. Transportation, warehousing, handling of materials, inventory management, purchase orders, market forecasting, repackaging, plant and warehouse locations, and customer service are all included in this category of operations thatsupport the efficient flow of commodities from producer to consumer.

- According to Philip Kotler, physical distribution entails organising, carrying out, and managing the physical movement of raw materials and finished items from the point of origin to the place of use in order to satisfy customer demands while making a profit.
- Actual distribution is described by William J. Stanton as "including the development and management of flow systems as well as the management of the physical flow of items."

Inventory control is therefore described as the management of both the real physiological flow of commodities and the flow system's administrative procedure. It is a strategy for managing the movement of the commodities.

Objectives of Physical Distribution: The two major objectives of physical distribution are maximising profit and ensuring consumer delight. A satisfied customer is a business's greatest asset. A company can satisfy its customers by making the proper number of the proper items available at the proper location and time, at the proper price. Customer satisfaction is increased by prompt and reliable distribution.

Additionally, the business may draw in more customers and increase profits by providing better service at a reduced cost for the product. This can be accomplished by increasing the productivity and effectiveness of physical distribution activities. By doing this, the company can save money, which will have an impact on the profit margin. Specifically, by reducing the costs associated with physical distribution, the profit position can be improved in order to maximise profits.

The other objectives of physical distribution can:

- Offer the proper products in the proper quantities at the proper times and locations at the lowest possible cost.
- Achieve faster shipping and a minimal inventory level.
- Set product prices through efficient physical distribution activity management.
- Gain a competitive edge over competition by providing superior customer service.

4.1.5.1 Strategic Importance

The strategic importance for physical distribution management includes the following:

(i) **Creating time and place utility:** Activities involving physical distribution support the creation of time and place usefulness. warehousing and transportation are used to accomplish this. The availability of the items at the precise location where they are needed thanks to the transportation system produces place utility. By keeping the items in storage and releasing them as needed, warehousing produces time utility.

(ii) **Benefits in reducing distribution cost:** The cost of physical distribution makes up a sizable portion of the product's cost. If these expenditures are managed methodically, the price of the product may go down. Cost savings may be easily achieved by careful and methodical planning of transportation timetables and routes, warehousing placement and maintenance, material handling, purchase orders, etc.

(iii) **Supports in stabilization of price:** Physical distribution aids in keeping costs steady. Even consumers anticipate pricing consistency over time. Utilizing warehouse and transportation resources effectively can help balance supply and demand, maintaining price stability.

(iv) **Improved consumer services:** Consumer service in distribution refers to making the appropriate number of items available at the appropriate time and location, i.e., where the customer requires.

4.1.5.2 Tasks in Physical Distribution Management

(i) **Materials handling:** It involves moving items into and out of a stock. It is made up of routine work that can be automated and standardised to be accomplished. Efficiency is increased by using digital information processing to control traffic, order picking, and conveyor systems. Because of innovative automated handling services and protective packaging, customer service has improved while physical distribution costs have fallen. Material handling and packing services have both sped up this process of orders and the transportation of consignments.

(ii) **Inventory planning and control:** The stock of goods that a company keeps on hand and is prepared to sell to clients is referred to as inventory. To quickly satisfy market demands, inventories are maintained. The chain between client orders and business production is represented by inventory. In actuality, inventory management serves as the hub for the complete physical distribution management. The key to success in the physical distribution game is effective inventory management.

(iii) **Order processing:** Order processing and inventory management are related. Providing outstanding customer service is viewed as being dependent on order processing. All aspects of receiving, recording, filling, and assembling products for delivery are covered. It must be reasonable and as short as possible from the time an order is accepted and the day when the things are shipped.

(iv) **Transportation:** A crucial component of physical distribution, it. It entails combining the benefits of each mode of transportation by using physical handling products and containers to allow transfers between various carriers. For instance, *"piggybacking"* refers to loading containers into railroad flat cars before loading them onto automobiles, while "flashbacking" refers to offloading containers onto water carriers. "*Air truck*" or *"birdy back"* refers to the interchange of containers among air and truck carriers.

(v) **Communications:** It is the transfer of knowledge and comprehension by one person to another. Included in this is the information system that need to connect producers, middlemen, and consumers. The channel's other participants can exchange information with each other more easily thanks to laptops, memory systems, display devices, and other communication technologies.

(vi) **Organizational structure:** To deliver the intended service to customers in the most effective way, the person in charge of physical distribution should coordinate all the actions into an effective system. Examples: How may the five physical distribution factors stated above be best coordinated to provide a team effort? What are some ways to prevent compartment formation thinking? (ii) Should the chief executive officer or the head of marketing be the ones to whom the central head in charge of all physical distribution operations reports if one is established?

Components of Physical Distribution:

1. **Order Processing:** The first step in every distribution operation is order processing. Order processing entails tasks like accepting the order, managing it, extending credit, billing, shipping, collecting payments, etc. Every customer anticipates that his purchase will be carried out promptly and in accordance with his specifications.

   As a result, processing orders is crucial. The order cycle time, or the time between when a consumer places an order and when the items arrive at their destination, should be maintained by marketers. The processing of orders should follow a standard method.

Physical Distribution with respect to pharmaceutical marketing in India: Placing the product at right place in right time is possible through physical distribution only. In pharmaceutical marketing. Physical distribution involves three main areas of concern as inventory control, warehousing and transportation.

Inventory control is a major area of concern to maintain low cost of distribution. Inventory the word includes raw material, packing material, work in progress and finished goods. So managing inventory level at minimum to meet customer need in time through minimum investment are the most important objectives of inventory management. Short supply or stock outs shouldn't be there because it is very harmful for the company in long run as the prescriptions get substituted.

In correlation with other department like sales, material, production and finance cost effec-tiveness can be achieved through best inventory control method. In Indian pharmaceutical industry, ABC analysis is most commonly used technique for inventory control to avoid short supply or stock out position.

In this ABC analysis, goods are organised according to their sales in decreasing order, and inventory is kept based on the top-selling item. One aspect is keeping inventory levels at a profit-making level, but it's also critical to make sure stockists and C&F agents get paid for their services.

The business often operates with a minimum quantity of inventory and cannot manage the payment delay. Additionally, the distribution manager must balance product supply and demand using the E.O.Q (Economic Order Quantity) approach. In addition, it's important to keep safety or buffer stocks on hand for emergencies. therefore inventory management and receivables are very important to maintain profitability trend in pharmaceutical industry.

Warehousing: It must adhere to the rules established by the Indian government and the GMP (Good Manufacturing Practices) standard as put down by the WHO (World Health Organization).

Many Indian pharmaceutical businesses own their own depots and C&F agents, which makes it simpler for them to reduce costs and make the most use of their resources. A good and ideal warehouse for storage is mostly dependent on its convenient location, accessibility to transportation, affordable labour cost, etc.

Transportation: Road transportation accounts for the majority of Indian pharmaceutical marketing since it is more affordable than air or rail delivery. Air travel for sterile preparations is favoured for some life-saving medications that need a precise temperature, such as injectable and parenteral medications.

4.2 Professional Sales Representative (PSR)

A pharmaceutical professional sales representative is either employed full time or part time in an organization, to promote and deliver the pharmaceutical products. They have given duties according to their profile and territory.

4.2.1 Duties of PSR

A PSR is a professional one whose role is given by a company to sell the product or service that they provide.

PSR serves as the main touchpoint between a company and its customers and makes sure that they are receiving the proper goods and services. It also finds new markets and consumer leads and pitches potential clients. The PSR serves as the firm's face in a variety of capacities, including taking calls and keeping an eye on the competition while fostering positive relationships with clients and looking for new business prospects.

Job responsibilities of PSR:

(i) Consistently make cold calls to decision makers in order to actively search for new business prospects.

(ii) Provides for the requirements of clients by selling goods.

(iii) Deliver compelling and expert executive-level presentations.

(iv) By planning and creating a timetable to contact present or potential retail stores as well as other trade components, one may serve current clients, get orders, and create new client relationships.

(v) By researching the kind of retail shop or trade aspect, v. modifies the substance of sales presentations.

(vi) Examines the number of existing and future dealers to focus sales efforts.

(vii) Submits orders using price lists and corporate material.

(viii) Provide management with regular activity and result reports, such as daily task summaries, weekly work plans, and monthly and annual territory evaluations.

(ix) Gathers information on pricing, goods, new items, delivery dates, and merchandising strategies from the existing market.

(x) Makes recommendations for changes to policies, services, and goods based on an analysis of performance and market trends.

(xi) Responds to customer complaints by investigating problems, finding solutions, producing reports, and offering management advice.

(xii) Maintains technical and specialised knowledge through participating in professional groups, attending classes, reading trade publications, and forming personal networks.

(xiii) Maintains records on local and client sales to provide historical records.

(xiv) Work with vendors and manufacturers on design and technical challenges.

(xv) Support new product efforts and take part in online marketing campaigns.

(xvi) Contributes to teamwork by completing pertinent tasks as needed.

Medical representative: A medical consultant is a PSR who works for a company to market and sell its goods, whether those goods be pharmaceuticals or medical devices.

A medical agent is the one who generates demand for pharmaceutical products and ensures their availability at stores and stockiest. They serve as the primary conduit between healthcare providers and pharmaceutical and medical firms.

Skills of a medical representative: While hiring medical representatives the employers look upon, his key expertise of good communication skills, interpersonal skills, verbal skills, negotiating skills, presentation skills. The candidate must have good planning and organizational skills and can deal with customer services. Must be confident and persistence with time-management skills for good sales and customer relationship.

Role of medical representative in a pharmaceutical company:

- A medical representative organizes appointments and meetings with the community and hospital based healthcare staff.
- He chooses and starts a new company based on sales performance.
- Showcasing and presenting pharmaceutical items to medical professionals including physicians, nurses, and pharmacists.
- Promotes the introduction of new items to the market and serves as a corporate representative.
- Pharma product promotion and market feedback about the product is also one of the responsibility.
- A medical representative cultivates and preserves cooperative working relationships with administrative support personnel and medical staff.
- Oversees the allocation of funds for conferences, catering, outside speakers, and corporate hospitality.
- Maintain thorough records of all interactions and the scope of subsequent reporting, and if at all feasible, surpass the company's yearly sales objectives. A medical representative Follow up on leads generated by the company.

Duties of medical representative: The three categories of a medical representative's responsibilities are as follows:

(i) **Prescription generation**: The most challenging task for a physician is to increase revenues, and the most efficient method to achieve this is to get doctors to prescribe the company's goods to patients they encounter.

(ii) **Customer coverage:** The medical representative must often meet with the physicians. To advertise the correct products to the appropriate physicians, a medical representative must be informed of the various specialisations of those specialists.

(iii) **Market intelligence**: A company's success or failure is determined by how well-equipped it is to meet the needs of its customers in light of the efforts of its rivals and the state of the market.

4.2.2 Purpose of Detailing

Detailing is essentially a crucial component of personal selling that is meticulously performed, learned, and rehearsed to the point that it seems natural when spoken aloud about the product. (Exactly like an actor would perform on stage or in a movie.)

Even though the script of the pharmaceutical product is written by someone else (usually the Product Manager), it looks as if, the medical representative has written the product detailing. Detailing is one of the most important tools in pharmaceutical marketing and advertising.

Pharmaceutical firms employ the marketing strategy of "**pharmaceutical detailing**" to inform doctors about certain pharmaceutical goods in the hopes that they would prescribe such medications more frequently. Detailing is a technique of prescribing the doctor through effective conversation to prescribe the product in one or more specific indications.

Purpose of Detailing

- Product advertisements for the business (a walking audio-visual).
- Develop and pique the doctor's interest in items that encourage the habit of prescription.
- Highlight the key benefits, features, and advantages of the items.
- Produce more prescriptions; this only depends on how well one can detail, or, to put it another way, how well one can speak.

Significance:

- Increases the current prescription of drug and sales volume.
- Helps to know the mentality of doctors.
- Provides latest drug information to doctors about the drug.
- Help the doctors to establish personal contacts with pharmaceutical companies.
- Enables the doctors to get free drug samples for trails of drug. These samples can be prescribed to the patients.
- Provides a way of releasing stress to the doctors.
- The gifts, letter pads, diaries calendars are also provided to doctors.
- It is an important means of employment to a large number of persons.
- Provide newer, effective and quality drugs to patients.

Steps of effective detailing: There are various directives of effective detailing in order of importance.

1. **Text**: very crucial because text is the only component of the message. Even a small variance might result in information that is inaccurate, misleading, or ambiguous. Misinformation might result from a minor addition or deletion. Perfection inspires interest and inspires confidence.

2. **Voice:** There are several facets to voice, including:
   - **Clarity of Speech**: Clarity gives the text a powerful impetus and keeps the detailing interesting. It makes other people more attentive.
   - **Audibility / Loudness**: Without being extremely loud or too quiet, the volume should be just right to be heard clearly in the examination room without making it difficult for the doctor to listen. Really, audibility ought to match the natural tone.
   - **Tone:** Tone is the emotions expressed in a voice. Words alone don't signify anything. The tone adds the emotions to the message that is conveyed, for example, a friendly tone. Accentuation and voice modulation.
   - **Modulation:** It is the voice's peaks and valleys. Voice modulation truly pulls out the message's emotions; without it, it would sound flat. It adds more force where it is needed, such as the brand name, the payoff line, the introductory sentence, etc. Emphasis provides the message the proper meaning. It must be applied properly. The concealed message is not revealed by excessive or insufficient usage. Finally, the confidence level is reflected in the modulation and emphasis.
   - **Timing and pause**: The doctor should be able to listen and understand due to the speed that the words are stated. Both too slow of a pace and too much speed makes it tough to grasp. A appropriate pause allows the doctor to process the information, and the right timing and tempo allow us time to reflect and regain our breath for new emphasis.
3. **Handling of Visual Aid/Folder:**
   - The doctor can read and see well when an oner or folder is handled properly. While handling this visual assistance, your first consideration should be the doctor's comfort.
   - A recent study found that the minimal seeing distance between the prospect's eye and the visual assistance is three feet. It offers complete clarity.
   - In order to prevent light reflection and vision distortion, a folder or other graphical images should indeed be carried at a 90 ° angle to the eye.
4. **Use of pointer:**
   - The synchronisation of words and the pointer helps the doctor focus on what is being presented and explained while also minimising distraction and personal planning. add up the punches that modulation and focus provide.
5. **Eye to Eye contact:**
   - Visual contact with the prospect represents the MR's degree of confidence and aids in determining whether or not doctors are truly interested in the opportunity.

- As you keep eye contact, many human aspects are involved. Doctors get intrigued by you, and they begin looking forward to your next communication.

6. **Body language**:
    - Non-verbal communication is responsible for more than 50% of communication, according to studies.
    - Your body's nonverbal communication, or body language, conveys a lot. The effectiveness of details is increased by a bold, skilled body language.
    - The customer's body language is equally significant. He or she is frequently providing feedback through his or her body language. To shape the details, it is essential to recognise these signals.
7. **Listening:**
    - Customers frequently voice concerns or offer comments, which makes listening a frequently overlooked yet crucial component of detailing. How the issue is handled will ultimately determine if we have been successful in gaining the physicians' trust. in advance of the prescription. You can only do this if you pay attention.
8. **Time Management**:
    - Doctors give Medical Representatives the amount of time they deserve, not the amount of time they need, so that they can finish their work.

Stages of detailing:

1. **Approaching:** According to Elmer Wheeler "*Your first 10 words are more important than the next ten thousand*". So, first approach to the doctor's cabin is very important and critical step. Some of the important instructions while approaching the doctor
    - Always take consent of doctors by sending the visiting card before entering.
    - Use good manners while entering, seating, talking and even leaving the doctors cabin.
    - Create such an atmosphere that doctors pay attention to the talk.
    - Start with the probing which means to put forward a question to doctors related to his patients or to his need. The question should be such that the doctors give an elaborate response.
    - Proper dressing sense and discipline of medical representative is helpful in impressing the doctor.
    - After this start the description about the product with the help of visual aids.
    - Medical representative should use impressive and enthusiastic voice tone while detailing.

2. **Development of interest**: The development of interest is very important for detailing success. Once, the medical representative knows about the needs of doctor, the detailing should be done accordingly. Usually the doctor is not interested in "our product" and "I" but only interested in his patients and his practice. So, medical representative should try to explain how the drug can provide benefits to patients.
3. **Use of product name**: While detailing the name of product must be mentioned as many times as possible by its "brand name" (and not by "our product" or "It"), because the main aim of medical representative is to visit the doctor to get the product prescribed by its brand name.
4. **Detailing the features:** When the doctor seems to be interested in medical talks, medical representative should start with major benefits, features, dose, route of administration etc. The benefits should be talked always before the features because the doctor is interested in patient benefits rather than product features. Example, a medical representative is performing detailing of an analgesic, to the doctor then the emphasis is on "*Potent Analgesia with drowsiness*". It means the product features are highlighted. On the other hand, if the medical representative says *"the patient will be able to pursue his normal routine"* the statement highlights the product benefits to patient. Such detaining will be more effective than first one.
5. **Sample distribution**: After describing the major benefits and features of product medical representative has to present the free samples of drug to doctor to initiate the trials of drug.
6. **Closing:** Closing of detailing is also very important. Medical representative has done a difficult job to attract the attention of doctor, created the interest, give details about product benefits and features, cleared the doubts if any and convinced the doctor. If at last, medical representative doesn't demand for prescription, his objectives will not be fulfilled. Example, Sir please extends your favor for product.... in terms of prescription.

   Medical representative can also request to doctor by adding some other personal statements like, Sir please prescribe ...... drug to your patients so that I can fulfill the sales target. My promotion is due next month and for that my sales target must be completed before due date.

Don'ts-In detailing:

1. Don't use samples and literature as the starting point of detailing.
2. Don't use superlatives in describing the product.
3. Don't use the same sales point every time, bring novelty in the discussion.
4. Don't make interruptions in detailing for this preplanning and studying of product related information is required, proper practice of study material is required.

5. Don't underestimate the doctor's knowledge, always keep updated knowledge of product.
6. Don't use our product or other general term for product. Always spell the brand name of product every time during detailing.
7. Don't hesitate to meet a doctor again and again because it is necessary for earning more prescription from doctor.
8. Don't criticize other product, company, medical representative or doctor.
9. Don't mention about competitor's brand.
10. Don't argue with doctor. The argument can be won but it will lose the prescribers. So disagree politely if it is must.
11. Don't maintain non-professional or professional relations with doctor. Try to make sound professional relationship. High personal intimacy can lead to indifference.
12. Don't influence the opinion of one doctor by mentioning the name of other prescribers.

Types of detailing:

1. **Personal detailing**: Medical representative personally meets the doctor and performs detailing about his company's product. The medical representative carries the free sample of drug, visual aids, printed materials about drugs and gifts sometimes with him. MR during detailing also clear the doubts of doctors related to drug. Thus he tries to impress the doctor for prescribing his product. He also tries to build sound professional relationship with doctor which is beneficial for future.
2. **Memorized detailing:** In this detailing the text material related to drug information is already printed by medical experts of companies. Medical representative is supposed to study the detailing thoroughly and then recite it in front of doctor. In this detailing no flexibility is shown while presenting the product, they have few merits and demerits as mentioned in Table 4.4.

Table 4.4 Merits and demerits of Memorized detailing

S.No	Merits	Demerits
1.	In this type of detailing all the important points about the drug product gets covered.	Seems to be artificial and lacks the normal vocabulary of medical representative. It is very monotonous.
2.	The details are prepared by professional experts at the head office of the company so free from the mistake.	Involves no interruption or doubts clearing during description phase of detailing.
3.	Less trained medical representative can also present such detailing to doctor.	Medical representative always acts as a monologue about the product.

3. **e-detailing:** This is most effective and modernly accepted method of detailing. It involves detailing the information related to drug using internet with few merits and demerits as mentioned in Table 4.5.

Table 4.5 Merits and demerits of e- detailing

S.No	Merits	Demerits
1.	Doctors having less time to attend medical representative can access the drug information online.	More money is required.
2.	Samples can also be ordered by doctor online.	Relationship building between doctor and companies becomes less
3.	A large bulk of information can be made available by means of e –detailing.	This has posed a threat to the job of medical representative.
4.	e-detailing has shown better prescription response over conventional detailing method in last 10 years.	It has lacks in clearing of doubts.

It can be better option that opted by most leading pharma companies as it uses a combination of e-detailing with medical representative detailing. This has increased their share of prescription in market and is able to create interest of doctors in detailing.

4.2.3 Selection and Training

Selection and training for employees in an organization are crucial part to run a company for longer period of time. Since, these are the fundamental scrutinizing process of personnel management.

A personnel management refers to the employees working in an organization and is concerned with effective control and proper use of manpower. It involves recruitment, selection, training and development of human resources.

A suitable selection from the personnel must be adhered to in order to develop and maintain an effective organisational structure, provide high morale and improved working relationships in a business. Through training or on-the-job education, people management creates the most individual and group growth. By providing a sufficient pay, promotion, or financial resources or bonus, this will acknowledge and meet individual demands and collective goals.

Areas under personal management:

A. **Recruitment:** It describes the process of identifying potential sources of the necessary employees and motivating suitable applicants to apply for positions inside the business.

Merit:

- It helps to provide the deserving personnel from which best are selected to fulfil the required standard.
- It is beneficial in the long run because not much training is required for potential candidates.

Sources of Recruitment:

Internal resources: The recruitment is done by promotion from within the organization. It is usually the best practice to fill jobs by promotion if possible. - Promotion, internal advertisement, departmental assessment, transfer, retirement, and employee recommendation

External resources: Management Consultants, Important To apply, Campus Recruitment, Local Newspaper Advertisements, Internal Advertisements, and Walk-In Interviews are the duty of the HR department or Personal department.

For this an advertisement is given in newspapers, magazines or by mail (Naukri.com, Times India job.com). The position is described in detail, and interested candidates must send their applications with complete biodata on the required form before the deadline. Typically, the application form asks for your full name, fathers name, age, address, educational qualification, Nationality, details of past experience if any etc.

Methods of recruitment:

Direct method: This method involves campus recruitment. The selected recruiter's team is sent to the technical institutes and interviews are arranged for the students.

Indirect Method: It entails placing ads in popular magazines, trade publications, or daily newspapers. The post's detailed description is provided, and candidates who are interested must submit applications and resumes.

Third party method: It includes the use of employment agencies, recruitment firms, consultants, friends or relatives to recruit the required personnel's.

B. Selection: After the recruitment the selection procedure is adopted for choosing the most deserving candidate for required post. It is done in following steps:

Scrutiny of application: The applications are received by indirect method and scrutinized. Incomplete applications are rejected.

Preliminary Interviews: Suitable securitized candidates are called for brief preliminary interview.

Selection test: Oral/written tests can be taken to judge I.Q or aptitude tests may be taken.

Selection interview: To determine if applicants are suitable, the recruitment committee and the applicant have a face-to-face conversation.

Physical Examination: It is done to make if the applicant is physically capable of doing the job. By sending job offers to the candidates who are deemed qualified in every way, positions are offered to them.

C. **Training and development:** The systematic process of enhancing an employee's knowledge and abilities for doing a certain job is known as training and development. Its major goal is to change new people's behaviour so they can do their jobs more effectively.

Types of training:

1. **Induction training:** it refers to training of the new employees. So that they are introduced with the existing staff and they are made aware of policies, procedures and rules of the organization.
2. **Promotional training**: if some employees are to be promoted, training is given to them to meet the requirement of the higher post.
3. **Refresher training**: From time to time, training is given to the employees to update their knowledge according to latest developments in their respective fields. E.g. Quality improvement programmes are held from time to time for the Pharmacy teachers.
4. **Safety training**: This training is given to workers to handle the sophisticated instruments to avoid any mishandling or fetal accidents. E.g. HPLC training is given to handle it properly.

Methods of training:

A. **On the Job:** This is the training given to the employees while working in the organization.

1. **Coaching and counseling**: It is given by supervisors to their subordinates. This is done by providing guidelines and instructions.
2. **Vestibule training**: It is given to the employees after their selection for the job.eg. MR are given formal training after their selection.
3. **Job rotation training**: This training is given when job rotation is done in the same organization.eg. Employees from tablet are shifted to parental section, then he requires training for the new responsibilities.
4. **Under study training**: This is provided to the employees who have been promoted to higher posts.
5. **Committee assignment**: The trainees are trained by forming committee. They are allowed to discuss on various job related topics. So that every member can share and learn new things about their job responsibility.

B. **Off the job Training:**

1. **Conferences and seminars:** There are the most common tool for off the job training. The companies conduct such seminars. Participants from various organizations and academic field are invited to present their papers.
2. **Lectures:** Guest lectures are conducted to share valuable knowledge and thoughts.

3. **Special courses:** Some big organizations conduct special courses so that the employees after passing these courses get useful updated information about their job responsibility.
4. **Case Study**: In this method, Written form is given to the employees and their views related to particular problems as well as its solution are taken in written. This helps in developing mental ability of employees.

4.2.4 Supervising

Today, working with staff to ensure sure they are aware of and able to carry out their job duties appropriately is often referred to as **supervising**.

The majority of the supervisory tasks that sales managers perform happen while they are dealing with new employees because of the autonomous aspect of the sales profession. It is a straightforward but possibly time-consuming operation that shouldn't be ignored because it is an essential part of a sales manager's duty.

Even though the term "supervision" is hardly used now, it was a crucial component of the manager's role when "management" was originally studied and it still is today!

People who supervise others carry out activities including watching, helping, giving advice, and receiving recommendations and criticism. One assists in addressing queries and handling objections while supervising.

Another tool for supervisors to use is customer relationship management (CRM), which allows them to keep tabs on a worker's daily activities, progress, and success during calls as well as how effectively they are utilising their time.

4.2.5 Norms for Customer Calls

Norms for effective customer service communication are based on following:

- Fill out forms and keep track of logs to construct accurate, thorough files for each client, giving you information about the target market and what they need from the business.
- Adhere to company policies to maintain high standards of service and make sure clients receive them.
- Encourage other salespeople and establish a team approach on the selling floor to raise spirits in the company.
- Report any necessary adjustments to procedures or equipment that will increase output and efficacy while reducing wasteful spending.

Techniques to provide customer service:

The easiest incentive strategy to use is this one. Your earning potential is based on the work you do and also how you do it. The idea is really straightforward. A performance-based incentive is more than just a commission, so keep that in mind. Fairness, transparency, and reachability are essential characteristics of incentives. The team will be expected to attain even more if the incentives increase in value as revenue statistics rise.

1. **Support customers as a team:** The most obvious motivational strategy is this one. The amount you will make depends on what you accomplish and how effectively you do it. It is a really basic idea. Remember that incentives for performance-based work go beyond just commission. Fairness, transparency, and reachability are requirements for the rewards. The staff will be inspired to work even harder by making the incentives more valuable as sales statistics rise.
2. **Listen to customers (and share their feedback):** Of all the incentive strategies, his is by far the most evident. Your income is based on what you are doing and how you do it. It is a fairly straightforward idea. Remember that performance-based incentives go beyond simple commission. The incentives must be reasonable, clear, and not difficult to attain. The staff will be inspired to work even harder if the incentives increase in value as revenue rise.
3. **Offer friendly for personable service:** Since customers are not machines, at the end it is how customer feel will matter the most. Customers are provided with services in a highly pleasant manner, essentially everything that supports the processes for selling the goods by adding a personal or friendly face.
4. **Be honest about what you don't know:** The truth is all that a consumer can ask for. You'll gain your customer's loyalty if you keep the lines of communication open and inform them at all times. It's OK for the customer service representatives to inform customers that they will contact the appropriate person and come back to them when they have an answer if they are unsure how to solve a problem. Keep an open line of communication with your clients and keep them fully informed to gain their respect and loyalty.
5. **Practice empathy:** Consider yourself the consumer, especially under challenging circumstances. Customers will value it, but it will also provide you a competitive edge. A corporation cannot succeed if its culture is one of passivity. Customers will appreciate empathy when it is demonstrated to them.
6. **Know your product:** The better a buyer can express interest in purchasing a thing, the more they will know about it. Some businesses need every new hire to attend boot camp and learn about the new product in order to guarantee they are familiar with it through and out. They also make sure to deliver the data for every new release.
7. **Remember that every second counts:** Customers despise waiting. They become more confident after receiving prompt assistance and having their

issues permanently resolved, at which point they are more inclined to maintain a connection with your company. Ensure that agents are encouraged to fix every issue fully at the same time; while speed is crucial, client happiness should never come before problem resolution timeframes.

8. **Improve as you go:** Of all the incentive strategies, this is the most evident. Your income is based on the work you do as well as how you do it. It is a really straightforward idea. Bare in mind that performance-based incentives go beyond just commission. Fairness, open communication, and reachability are requirements for the rewards. The team will be inspired to accomplish even more if the incentives increase in value as revenue statistics rise.

4.2.6 Motivating

The Latin term "motivation" means "to move." An inner condition known as a motivation can energise, activate, direct, or channel behaviour toward a certain purpose. The process of motivating individuals to work toward achieving organisational goals involves establishing motivating organisational settings. In order to motivate individuals to work with enthusiasm, devotion, and a sense of responsibility, certain circumstances must be established.. Numerous aspects of the organisation are influenced and influenced by motivation, which is a widespread notion. It allows managers to comprehend the motivations behind people's actions.

The most obvious motivational strategy is this one. The amount you will make depends on how you do it and also how you do it. It is a really basic idea. Remember that incentives for performance-based work go beyond just commission. Fairness, transparency, and reachability are requirements for the rewards. The staff will be inspired to work even harder by making the incentives more valuable as sales statistics rise.Of all the incentive strategies, he is the most evident. Your income is based on what you perform and on how effectively you do it. It is a fairly straightforward idea. Remember that performance-based incentives go beyond simple commission. The incentives must be reasonable, clear, and not difficult to attain. The staff will be inspired to work even harder if the incentives improve in value as revenue rise.

1. **Implement performance-based incentives**: Of all the incentive strategies, this one is the most obvious. Your income is based on what you accomplish and how effectively you do it. It is a fairly straightforward idea. Remember that performance-based incentives go beyond just commission. The incentives must be reasonable, clear, and not difficult to attain. The team will be inspired to work even harder if the incentives value increases as sales volume rises.

2. **Provide regular and useful sales meetings:** You inspire your team as a marketing manager by being available to them. It also contributes to the development of trust among you and the team members. Praise good work

whenever you have the chance to. Make it known to your staff that you are available to them and have an open-door policy. Set an example. Train team members to acquire useful skills and give up harmful habits. Salute a successful job. Never publicly criticise or humiliate employees of your sales team; instead, speak with the person directly if there is an issue. It becomes simple to understand what drives your sales team after you gain their trust.

3. **Be a leader, a coach and a cheerleader for your sales team:** Being accessible to your team as a sales rep inspires them. Additionally, it promotes trust among team members and you. Whenever you have the chance, give credit where credit is due. Tell your employees that you are available to them and that you follow a policy of open doors. Set a good example. The team members should be coached to assist them acquire useful skills and break harmful behaviours. Applaud a successful job. Never publicly chastise or humiliate a member of your sales force if there is an issue; instead, speak with the offender. Learning what drives your sales team is simple once you gain their trust.
4. **Use non-cash rewards to encourage your sales crew:** There are several non-cash methods for inspiring a sales force. These rewards are inexpensive and can be used frequently.
   - Invite sales representatives to marketing or planning sessions.
   - As a reward, grant an extra day of vacation.
   - Grant easy access to senior staff or the firm CEO.
   - Distribute local gift cards for objectives other than meeting sales targets.

   You ought to think creatively in this situation. Additionally, by knowing and understanding your sales staff, you may discover what drives each member personally, enabling you to offer tailored incentives.
5. **Provide opportunities for professional development:** You are achieving a lot by giving your sales personnel opportunities for professional growth and training.

   Your sales staff will feel appreciated, you'll raise their level of expertise, they won't become complacent in their selling methods, and they'll be more loyal to you and the business.

   The secret to achieving and sustaining sales success is to inspire the sales staff. If you can inspire your team to perform, you have uncovered the key to developing wonderful sales professionals that can achieve and persist with tremendous accomplishment.

4.2.7 Evaluating

Evaluating the personnel's: Once the employee has been selected and trained, evaluation of his performance is done in terms of requirement of the post.

Procedure of evaluation: It involves following steps

- Clear cut performance standards must be set against which employee can be evaluated.
- Employees are informed of performance expectations,
- periodically the actual performance is measured.
- Compare the actual performance with the set standards.
- The positive and negative points of comparison are discussed with the employee either orally or in written. So that employees are aware about his strength and weakness.
- Initiate corrective actions if necessary, such as counselling or corrective of weak points which is a slow but effective process.
- Every selling manager is tasked with the ongoing, informal responsibility of reviewing and evaluating his salesman. However, informal appraisal is insufficient to arrive at a genuine and insightful estimate of a salesman's value in either absolute or comparative terms. Any assessment program's ultimate objective is to increase a salesperson's value to the business. The evaluation procedure must include the following in order to do this:
    (a) An examination of the salesman's abilities, habits, aptitudes, and attitudes
    (b) An examination of his sales history, including his attempts and successes.
    (c) A description of the course that the development function will follow.

Your sales team will be sufficiently motivated by the correct plan to assist you in achieving your overall business objectives without jeopardising your profitability.

There isn't, however, a compensation strategy that works for everyone. The industry, company size, duration of the sales cycle, and other considerations must all be taken into account by the business owner when determining how much to pay salespeople. The various alternatives you might want to take into consideration are broken down below to assist you make sure you create a compensation plan that works best for your business and your sales department. corporate profit. In addition to helping the sales manager assess the performance of his sales force and boost efficiency, appraisal and performance assessment systems have also been proved to be helpful in:

(a) using salesmanship as a method of persuasion.
(b) Salesperson motivation and managerial leadership.
(c) Determining the salesforce's requirement for ongoing training and development.
(d) Improving sales tools such working papers, demonstration materials, etc.

(e) Determining and reorganising the job assignments and territory for salespeople.

(f) Improving sales strategy, such as by scheduling phone cycles, visits, and work preparation.

(g) Implementing sensible incentive and compensation programmes that are complemented by a national evaluation process.

4.2.8 Compensation and Future Prospects of the PSR

Compensation to the PSR: Compensation refers to the salary and other benefits such as house rent allowances (HRA), Free residential accommodation, City compensatory allowance (CCA), medical allowances, group insurance scheme, leave travel concession (LTC), travelling allowances (TA), Bonus, Retirement benefits etc.

Adequate compensation helps in retaining good employees and encourages them to work with full potential. It helps in motivating the employees. Employees who are working well and giving good results can be given extra compensation in terms of bonus, which will motivate them as well as other employees to perform their best. Thus adequate compensation leads to improvement in performance of the employees.

Following points should be taken care of during preparation of compensation plans:

Adequacy: The amount of compensation should be in proportion to the job responsibility.

Simplicity: The compensation structure should be straightforward so that employees can understand it.

Flexibility: It should be flexible enough to operate efficiently throughout the year (So that anytime compensation can be given to the employees throughout the years).

Promotion: A provision should be made in the compensation plan to provide promotion in post and salary for continuous long and devoted service of the employee.

Choosing a sales compensation plan is an important decision to make for any organization. The ideal strategy will effectively inspire your sales team to support you in achieving your overall business objectives without jeopardising your profitability.

There isn't a compensation strategy that works for everyone, though. When selecting how to pay sales staff, each business owner must take into account a variety of elements, including the industry, the size of the firm, the duration of the sales cycle, and others.

Here is a description of the various choices you might want to think about in order to make sure that you create a reward plan that works best for your business and your sales staff.

1. **Straight Salary:** Funding sales pay systems do exist in some businesses, despite their rarity. Even while it may be competitive, under this type of agreement, you would simply pay your salesperson a fixed wage, just like the rest of your team. There aren't any commissions, incentives, or other sales incentives.

   This type of compensation package is typically employed when your sector prohibits direct sales, when your sales staff is tiny or your salespeople work together or small groups where everyone participates equally, or when you anticipate that your salespeople will spend a significant amount of their time on tasks other than selling.

   However, since there are no rewards, these programmes rarely serve to motivate salespeople.

2. **Salary plus Commission:** The most common plans in use right now for sales compensation may include wages and commissions. They are designed such that lower base pay for salespeople is offset by commission payments, which make up the majority of their overall compensation.

   Organizations use conventional committee sales compensation programmes when there are opportunities to help all salespeople with this structure and when there are suitable procedures for portraying the data to ensure that the splits are fair and accurate.

   This sort of plan is generally preferable to a straight salary since it incentivizes employees to put in more time and produce more results. Additionally, it offers more consistency because salesmen will still be paid even if they aren't making any learning, when sales are slow during particular months, or if the market becomes unstable. But administering it might be more difficult.

3. **Commission Only:** Plans for committee sales pay are precisely what they seem to be; salespeople only receive remuneration for the goods they produce. There is no guarantee of financial gain.

   These kinds of programmes, which are based only on sales achieved, are simpler to manage than salary plus incentive and offer more value for your money. They also tend to attract fewer applicants, but they do attract the greatest and hardest-working sales people who are conscious of the possibility of high pay since they are adept at closing transactions. However, they can also encourage aggressive behaviour and a lack of financial stability in your sales team members, which can lead to a high turnover and strain burnout in sales representatives.

4. **Territory Volume:** The majority of companies with collaborative corporate cultures utilise territory volume sales remuneration packages. At the

conclusion of a reimbursement period, researchers work through the territorial volume calculation. The total revenue generated by the region is then divided evenly among all the sales reps that worked there. This technique works best when your sales territory are affluent enough to offer competitive remuneration, your sales team supports one another in achieving common goals, and your sales areas are well defined.

5. **Profit Margin:** We also have compensation plans for profit margin sales. These programmes base salespeople's compensation on the way the business is doing. Profit margin strategies are typically used by startups who need cash. The profitability plan should be utilised if you are certain that your salesmen can make ends meet during difficult times, if you can also add long-term incentives like equity shares, and if you have several incentives and benefits to tempt salespeople, such as flex time.

Future prospects of professional sales representative: In future an experienced professional sales representative can opt for the role of sales manager, regional sales manager and sales consultant. But, priorities for the future sales representative would be following:

- He/ She should be able to use data to understand what customers want (and how to give it to them).
- He/ She should be a technical expert to be more helpful.
- He/ she should be able to cultivate cross - functional connection.
- He/ she should be able to connect with the customer's emotions.

Answer the Following Questions

Q1. Define channel design. What are the steps involved in designing channel?

Q2. How do you evaluate Channel Member Performance?

Q3. What are channel members? Include the primary responsibilities of channel participants in channel distribution.

Q4. Enumerate the decisions taken by marketers to decide a particular channel.

Q5. Write the classification of distribution channel in marketing.

Q6. Write the functions of wholesaler in marketing.

Q7. Define retailer and enlist its functions.

Q8. Write a note on chain stores with its merits and demerits.

Q9. Define Distributor or Super Stockists or Super Distributor

Q10. Define Stockists, Sub-Stockists, Clearing and Forwarding (C&F) Agent.

Q11. Enlist the various functions of super distributor in pharmaceutical marketing.

Q12. What is the role of retailers in pharmaceutical marketing?

Q13. Define detailing and explain its significance.

Q14. Write a note on Job Profile of a Medical Representative.

Q15. Summarize the importance of motivation in marketing.

Q16. Differentiate between horizontal and vertical conflicts.

Q17. Elaborate the evaluation parameters of personnel's their compensation

Q18. Write note on types and methods of training process.

Q19. Explain various components of physical distribution.

Q.20. Classify distribution channel. Enlist the functions of retailer and wholesalers.

MCQs

1. Which component have the ability to change channel member's behavior.
   A. Channel power B. exerting force
   C. Persuasiveness D. Authentic authority

2. The retail supply chain does not include.
   A. Manufacturer B. Wholesaler
   C. Retailer D. Regulators

3. Distribution is the process of a producer putting their services or goods in as many retail locations as they can.
   A. Selective B. Integrated
   C. Intensive D. Multi-channel

4. Staffing needs
   A. Man power planning B. Communication
   C. Authority D. Coordination

5. The first step in planning is ....
   A. the setting of goals B. Creating new spaces
   C. Finding substitutes D. Follow up

6. Poor selection will result in more costs for.............. and oversight.
   A. Workplace quality B. Recruitment
   C. Training D. None of the aforementioned

7. Which of the following external factors has the biggest impact on hiring?
   A. Sons of the soil,
   B. the labour market,
   C. the unemployment rate, and
   D. supply and demand are just a few examples.
8. Which of the following notions is connected to improving employee productivity?
   A. scription
   B. stereotyped danger
   C. inspiration
   D. need
9. A channel for marketing is:
   A. A network of firms working together to make a good or service available for use or consumption.
   B. The process of selling goods or services.
   C. A strategy for spreading advertising.
   D. A procedure for giving the manufacturer customer feedback.
10. The average waiting time of customers to receive receipts of goods bought are said to be
    A. Jobber time
    B. Spatial time
    C. Delivery time
    D. Lot time
11. The scenario is classified as
    A. network trade release
    B. channel distribution rights
    C. channel ordination
    D. channel conflict when one channel member's actions prevent another message board from attaining its aims.
12. Is the employee's compensation a reward for them?
    A. Performance
    B. Work
    C. Organizational Contribution
    D. Intelligence
13. Who makes decisions regarding compensation?
    A. The HR Unit and Manager
    B. The Worker
    C. The Layperson
    D. The BOD

14. Job appraisal is a process, right?
    A. one-time process
    B. annual process (at the conclusion)
    C. ongoing
    D. random
15. Benefits are one form of compensation.
    A. Monetary,
    B. non-monetary,
    C. both "A" and "B," and
    D. none of the above
16. For evaluating channel member performance which one is odd
    A. Market variable
    B. Product variable
    C. Company variable
    D. Price variable
17. Which market factors do not fall within the fundamental subcategories?
    A. Geographical market
    B. Market size
    C. Market density
    D. Market for distribution
18. determining channel design are the most crucial corporate factors.
    A. Goals and plans
    B. Financial capability
    C. Managerial skills
    D. All
19. Abbreviations for CRM
    A. Customer relationship management,
    B. customer relationship manager,
    C. cost relationship management, and
    D. none.
20. The technique known as can be used to evaluate channels during the design phase.
    A. SCPCA
    B. SPCCA
    C. CSPCA
    D. ACCPS
21. The channel conflict can be classified majorly into .... categories
    A. 4
    B. 5
    C. 6
    D. 8
22. ....conflict occurs among firms at the same level of the channel
    A. Horizontal channel
    B. Vertical channel
    C. Multi channel
    D. Inter type channel
23. The degree to which a conflict is deemed urgent or requires the channel leader's attention is referred to as its
    A. Conflict magnitude
    B. Conflict energy
    C. Conflict power
    D. Conflict standard

24. ABC analysis is used for
    A. Conflict management
    B. Inventory management
    C. Channel management
    D. Training management
25. In EOQ "E" can be expressed as
    A. Economic
    B. Evaluation
    C. Efficient
    D. Equal
26. There are .... commandments of effective detailing in order of importance
    A. 9
    B. 8
    C. 7
    D. 6
27. Among the given which is unmatched
    A. Induction training
    B. Promotional training
    C. Safety training
    D. Departmental training
28. Direct marketing channel is
    A. Channel1
    B. Channel2
    C. Channel3
    D. Channel 4
29. With respect to wholesalers which one is not correct?
    A. Manufacturer wholesaler
    B. Retailer wholesaler
    C. Distributor
    D. Broker
30. V.P.P can be expressed as
    A. Value payable post
    B. Variety payable post
    C. Value payable percentage
    D. Value per post

Answer key to MCQs

1. A, 2. D, 3. D, 4. A, 5. A, 6. A, 7. D, 8. C, 9. A, 10. D,
11. D, 12. C, 13. A, 14. C, 15. C, 16. D, 17. D, 18. D, 19. A, 20. A,
21. A, 22. A, 23. A, 24. B, 25. A, 26. A, 27. D, 28. A, 29. D, 30. A

CHAPTER 5

Pricing

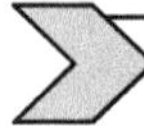

5.1 Meaning of Price

Price refers to the sum paid for a product, service, or concept. Without a price, there would be no marketing in society since a product, service, or idea is exchanged or offered for sale regardless of its value or worth to the potential customer.

Pricing is the art of expressing to clients the worth of a product or a unit of a service at a particular time in numerical units (such as rupees or dollars).This is a technique that a firm may use to determine the price at which it will offer its goods and services. The cost of production, market conditions, competition, brand, and product quality are all factors that the company will consider when determining prices for its products.

5.1.1 Importance of Pricing in Marketing

Pricing, being one of the 4 Ps of the marketing mix along with product, promotion, and place, is a crucial component of financial modelling. The four Ps all have cost centers except for price, which generates income.

1. When combined with marketing, price becomes a powerful instrument for persuading customers to purchase things. To enhance sales, it captures customers' interest and emphasizes the brand's image. Sometimes businesses maintain a steady pricing focused on recovering costs at a specific percentage while concentrating on other marketing mix components.
2. Setting rules and bounds for management to develop marketing plans through the finalization of price in conjunction with other marketing mix elements.
3. Standard of living is also influenced by pricing. Consumers have more purchasing power the more affordable things are in the economy. Price reflects the market's purchasing power.
4. Price is a potent tool against rivals.
5. Sales profits are based on price. Organizations frequently use this freedom for defensive or offensive pricing tactics since it is the most versatile of the marketing mix elements.

6. Price has an impact on two different management choices. Price-setting for new products comes first, followed by price adjustments for already-available products based on factors such as market conditions, costs, etc.
7. Organizations employ price in a variety of ways depending on the marketing programme, including demand-oriented strategies, cost-oriented strategies, competition-oriented strategies, and ethical considerations.
8. Pricing should be carefully chosen in relation to the other elements of the marketing mix. The market for a different product in the same manufacturer's product line is impacted by a product's price. For instance, the sales of two soaps from the same manufacturer that are priced similarly but have distinct attributes will be affected, and the client may find it difficult to decide which soap to buy. Prices should be set in accordance with the features of the product and should be accompanied by vigorous promotional efforts like sales, feature education, etc.
9. In addition to communicating the product's quality, price should be established in accordance to the provided value and perceived value of the good. Customers might assume a product is of inferior quality if it is priced extremely lowly while communicating its features better than the competition. In these situations, firms must make significant promotional investments and communicate properly while stressing the services offered, such as warranties and brand value.
10. Prices ought to be decided upon after consulting with wholesalers. Most businesses want to give distributors better profit margins since they aggressively market products to consumers by acting as wholesalers and retailers.
11. High promotional efforts incur costs. Businesses must base their pricing decisions on the costs of advertising, public relations, etc. The companies must carefully examine the promotional expenditure and evaluate whether it will lead to scale economies in both production and marketing. This will lower the unit cost and give flexibility to modify prices.

5.1.2 The Goals of Pricing

- Pricing establishes a company's financial objectives (i.e., profitability) by focusing on the rate of return on investment.
- Price should support a product's positioning in the market, be consistent with the other components of the marketing mix, and be reasonable in light of the circumstances.
- It also meets the objective of profit growth, establishing market position and maintaining market share to promote new products by attracting the customers etc.
- Increases consumer interest in the product to increase an organization's long-term revenues.

- Finally leads to maintain price leadership of product and quality, thus providing stability of the organization in terms of price.

5.1.3 Determinants of Price

The main factors that affect a product's pricing are:

1. **Cost of production:** If the price of the product is not set so that it covers all of the costs of manufacturing, then all production operations will have to be discontinued.
2. **Profit-margin desired:** To ensure profitable sales, the product's price should include a fair (or intended) profit margin.
3. **Competitor's pricing:** In the current atmosphere of fierce competition in marketing, no businessman could ignore the pricing tactics employed by competitors when determining the price for his own items. In any scenario, a manufacturer's price for a product cannot be much more or lower than what its rivals are charging for identical products.
4. **Government's policy of price-control:** In this regard, a manufacturer's pricing plan must adhere to legal requirements where, under specific conditions, the government has established maximum retail prices.
5. **Consumers' buying capacity:** A product must be developed to satisfy the needs and preferences of the target market, and its price must be established so that they can afford it, per the current marketing philosophy. Otherwise, they could not find the offering appealing, which would make selling the goods a "huge" challenge.
6. **Product-life cycle stage:** The maker must consider the specific stage of the product's life cycle that the product is in when determining its price. As an illustration, the product's price may be kept low throughout the introduction phase, then slightly raised during the growth phase, and then dropped once again at the point of saturation.
7. **Demand-supply conditions:** Whether a product should have a high or low price will largely depend on the supply-demand dynamics that apply to that particular product. Even a high price may be effective if demand exceeds supply. On the other hand, when supply is limited and demand is high, only a low price will draw customers.
8. **Organizational decision making and implementation**: The management's abilities and sound judgement play a significant role in successful pricing. To create a successful pricing plan, the upper management and lower management should collaborate. For the information to flow from customers and distributors to all of the organization's interested personnel, the proper systems must be implemented. Effective cost, demand, and competition strategy analysis is

necessary before making a pricing decision. The company must ensure that the appropriate personnel are performing the appropriate duties at the appropriate times.

9. **Product differentiation**: How unique is the product in comparison to other products on the market? The distinction can be seen in terms of its characteristics, placement, design, shape, etc. Depending on how the customer feels about the goods, there is a difference. The business determines a price based on the product's value to the client and its uniqueness. As an illustration, HUL (Hindustan Unilever) sells bath soaps that are priced differently based on the distinctiveness of each customer group.

10. **Target market attractiveness and economy:** The price strategy is also influenced by the target market's purchasing power and consumer demographics (early adopters, laggards, etc.). If the target market's economy is strong, there are lots of opportunities for the company to increase sales through various pricing techniques and tactics, such as market penetration, market skimming, perceived value pricing, demand differential pricing, etc. Prices are typically low when the economy is poor. In these cases, there are frequently no competitors, but the pricing is still determined to meet consumer demand. If prices are set too high, a rival will typically enter the market with a product that is less expensive.

5.1.4 Pricing Techniques and Tactics

5.1.4.1 Pricing Methods

Cost-oriented method and market-oriented method are the two main groups.

A. **Cost-oriented method:** Because cost serves as the foundation for a potential price range, some firms may think about using cost-oriented tactics to determine pricing. Listed below are some instances of cost-based pricing and methodologies:

Cost plus pricing: Cost must be raised by a preset proportion in order to fix the price. For instance, the selling price will be Rs. 220 if the product costs Rs. 200 per unit and the marketer anticipates a 10% cost margin. Profit is defined as the difference between the selling price and the cost. This technique is easier since marketers can calculate the costs more rapidly and increase the selling price by a specific amount.

***Mark-up pricing*:** Mark-up pricing is a variation of cost pricing. Instead of a proportion of the cost price, markups are calculated here as a percentage of the selling price. Businesses that use cost-effective methods use markup pricing.

***Break-even pricing*:** The company calculates the volume of sales required in this case to pay all the fixed and variable expenses. The price at which sales revenue and product costs are equal is known as the break-even point.

In other words, neither a gain nor a loss has occurred. The company needs to sell 40,000 units in order to break even if the fixed cost is Rs. 2,00,000, the variable cost per unit is Rs. 10, and the selling price is Rs. To turn a profit, the corporation needs to sell more than 40,000 units. If the company is unable to sell 40,000 limits, the selling price will have to be raised. The break-even point is calculated using the formula below:

Contribution = Selling price – Variable cost per unit

***Target return pricing*:** The corporation in this instance sets its prices to get a specific level of return on investment. The formula below can be used to determine the target return price:

Target return price = Total costs + (Desired % ROI investment)/ Total sales in units

Example: The desired return price will be Rs. 7 per unit as shown below if the entire investment is Rs. 10,000, the targeted ROI is 20%, the total cost is Rs. 5000, and the expected total sales are 1,000 units:

Target return price: 5000 + (20% × 10,000)/ 7000 = 7

This strategy has the disadvantage of deriving prices from costs without taking into account market factors like competition, demand, and consumers' perceptions of value, like other cost-oriented systems do. However, this approach ensures that prices are higher than all costs and generate profits.

Early cash recovery pricing: When market predictions indicate that the market's estimated lifetime is expected to be brief, few enterprises will establish a price to return their investment early, as is the case with products with a strong fashion or technology component. Such pricing can also be employed when a company thinks a major competitor with lower prices will soon enter the market and drive out smaller players. In these circumstances, businesses may select a pricing level to increase short-term profitability and minimize medium-term risk to the business.

B. Market-oriented Methods:

Perceived value pricing: Numerous businesses base their pricing on how valuable they perceive their consumers to be. They view customer perception of value as the main determinant of price, with the company's costs coming in second. A few of the many factors that might influence the customer's perception are advertising, sales strategies, an effective sales team, and after-sale service staff. By paying a higher fixed price, customers will perceive a higher value, and vice versa. To establish the customers' perceived value as a benchmark for appropriate price, market research is required.

***Going-rate pricing*:** The pricing standard in this instance is the price determined by the primary rivals. Independent of their costs or demand, smaller businesses may change their prices if a major competitor does. There are three additional ways to divide going-rate pricing:

- **Competitors 'parity method**: A business may choose to match the price of its main rival.
- **Premium pricing**: If a company's products have more special characteristics than its main competitors, it may charge slightly more.
- **Discount pricing:** If the company's products don't have certain features compared to those of its top competitors, it will maintain a lower price.

The going-rate strategy is well-liked since it tends to lessen the possibility of price wars breaking out in the industry. Additionally, it reflects the industry's common understanding of the cost at which a respectable return would be generated.

Sealed-bid pricing: Large purchases or contracts, particularly those from businesses or government agencies, are subject to this price. In response to an advertisement, the businesses offer sealed bids for the jobs. In this scenario, while the buyer seeks the lowest price possible, the seller is supposed to provide the best quotation or tender. A company must submit a lower price bid in order to win the contract. To achieve this, the company must determine the price offer by anticipating the competitors' pricing strategies.

***Differentiated pricing*:** For the same good or service, different businesses may charge different prices. Here are a few examples of differentiated pricing:

- **Customer segment pricing**: In this method, different client groups are charged variable rates for the same item or service depending on the size of the purchase, the conditions of payment, and other elements.
- **Time pricing**: By doing this, it will be possible to preserve the various prices that are charged for the same commodity or service throughout the year. It includes off-peak pricing, whereby costs are reduced at off-peak times or seasons of low demand.
- **Area pricing:** Here, the same product is sold at several price points in various market regions. As an example, a company might provide reduced price in a new market to attract customers.
- **Product form pricing:** In this instance, prices vary according to the numerous product types, but they are not fairly based on their unique costs. This pricing strategy, for instance, is applied to soft drinks that are 200, 300, 500 ml, etc.

5.1.4.1.1 Pricing methods of inventories

The pricing of inventories is very critical because it should generate profits, indicate the current market price as well as retain the customer satisfaction.

There are different methods for pricing of inventories:

1. **First in First out method (FIFO)**: This strategy involves selling or using the goods that were first received. Until it runs out, the item is priced at the cost of the oldest consignments. The issues are priced at the cost of the following lot, such as if:

   March 10th ----- 300 units purchased at Rs. 5 per unit

   April 5th -------- 500 units purchased at Rs. 6 per unit

   If on April 25th ------ 400 units have to be priced for selling

   Then Price for 300 units = 300 × 5 = 1500

   Price for 100 units = 100 × 6=600

   So, Total price for 400 units = 1500 + 600

   = Rs.2100

2. **Last in first out (LIFO) Method**: In this procedure, the consignment is initially sold and the proceeds are utilized to determine the stock's worth until the entire consignment has been used up. When it is exhausted the next consignment is used for pricing and so on, for example, if:

   October 24th ----- 500 units purchased at Rs. 10 per unit

   November 2nd -------- 300 units purchased at Rs.12 per unit

   If on November 18th ------ 600 units have to be sold

   Then,

   Price for first 300 units = 300 × 12 = Rs. 3600

   Price for next 300 units = 300 × 10 = Rs. 3000

   So, Total price for 600 units = 3600 + 3000

   = Rs.6600

3. **Highest in first out (HIFO) Method**: This strategy uses the materials lot that earned the highest price to sell it first and utilize the proceeds as the basis for the stock price. For instance;

   April 8th ----- 200 units purchased at Rs. 6 per unit

   April 15th -------- 300 units purchased at Rs. 8 per unit

   So, On April 28th ------ 400 units have to be priced

   Price for 400 units = 400 × 8

   = Rs. 3200

4. **Average cost Method:** When a fresh stock of goods is received, the total value of the items in stock is multiplied by the number of units to arrive at the average price per unit. Until the next consignment is received, the entire inventory of goods will be sold at this price. The new average is once more calculated upon receipt of a new shipment, and so on. For instance, if:

March 12th ----- 800 units purchased at Rs. 15 per unit

March 25th ----- 1000 units purchased at Rs. 20 per unit

If, on April 7th ------ 1500 units have to be sold

Then, the average cost per unit = 20+15/2 = 17.5

So, Price for 1500 units = 1500 × 17.5 = Rs. 26250

5. **Base stock price method:** With this method, a fixed price is always placed on a minimum amount of stock that is kept on hand as a reserve. Only in an emergency is the base stock, also known as the minimum stock or safety stock, used. The value of this stock is fixed (The price at which it was purchased plus the other expenses like storing, transportation etc.). The stock above this stock is priced using a different methodology (LIFO, FIFO, or HIFO).

6. **Replacement (or market) Price method**: The materials are priced using this approach at the going rate on the day of issue. As a result, every time the material lot is provided, the replacement price is computed.

7. **Standard price method**: The material lot is priced at a standard predetermined or pre-estimated price. This standard price is decided initially by summing up the cost of material plus labor and transportation charge and overhead expenses like storage charge, inspection cost, taxes etc. This price is used for pricing all this stock and it is applicable for definite period such as, quarter of year, six months or for a year depending on the change in market situations. The standard price should be revised after this period is over.

8. **Inflated price method**: This fee is only applied to products like glass equipment and volatile ingredients that are prone to waste. For instance, if:

   100 packets of methanol were purchased at Rs. 80 per packet.

   There is wastage of 10% of material during storage.

   So, the material left after wastage =100-10=90

   Price = Total price/quantity of material left

   = 80 × 100/90

   Hence, Price = 8000/90

   = Rs. 88.88 or Rs. 89

9. **Actual Pricing method**: It is used only in those cases if original materials are very costly and some customers can't afford them. In such cases substandard materials are purchased to make products and price at low costs to meet satisfaction of specific customers. The cost of substandard materials is lower than standard original materials.

10. **Weighted Average method**: The resources are priced using this technique at their weighted average, as if:

    On June 10th----- 500 units purchased at Rs. 25 per unit

    On June 18th ---- 600 units purchased at Rs. 30 per unit

    If, on June 25th ------ 800 units have to be priced for selling

    So, Weighted average will be calculated as:

Unit cost	Weight	Weighted cost
25	500	12500
30	600	18000
Total	1100	30500

Therefore, Weighted average cost = 30500/1100

Weighted Average cost = Rs. 27.72

So, Price for 800 units will be at the rate of Rs. 27.72 per unit

Hence, Price = 800 × 27.72

= Rs. 22181.82

5.1.4.2 Pricing Strategies

1. **Operations-oriented pricing:** When matching supply and demand through altering prices, maximizing productive capacity, or achieving operational efficiencies are the goals.
2. **Revenue-oriented pricing:** Regardless of whether the marketer seeks to maximize profits (i.e., the excess income over costs) or merely to cover costs and break even, it is often referred to as profit-oriented pricing or cost-based pricing. As an illustration, take a look at dynamic pricing, a type of revenue-oriented pricing commonly known as yield management.
3. **Customer-oriented pricing:** When attracting more consumers, fostering cross-selling opportunities, or taking into account the varied degrees of client purchasing power.
4. **Value-based pricing:** When a business uses prices to reflect market value or links prices to the intended full appreciation in the eyes of the customer, a phenomenon known as "image-based pricing" occurs. Value-based pricing, an illustration of a premium pricing stance, aims to uphold or pursue a rich image by supporting the entire positioning plan.
5. **Relationship-oriented pricing:** When a marketer establishes prices to create or maintain relationships with current or prospective customers.
6. **Socially-oriented pricing:** When a specific social attitude or conduct is what's being promoted or avoided. Consider how high cigarette taxes might discourage smoking as an example.
7. **A time-based pricing strategy**: Businesses that offer seasonal goods or services or last-minute purchases tend to use it.

8. **Cost-plus pricing:** In order to establish how much to charge for goods and services, this pricing strategy takes a cost-based approach. The price of raw materials, the price of manufacturing, and the cost of overhead are added together when determining the cost-plus pricing of a service or an item. Your cost-plus price is the result of adding a markup percentage (your profit margin) to this amount. You will always operate at a profit as long as all costs and sales have been accurately predicted.
9. **Discount pricing:** A pricing approach that charges less for goods and services. Discount prices can take the shape of holiday sales, loyalty discounts, etc.

5.1.4.2.1 New-Product Pricing Strategies

(i) **Market-skimming pricing:** By boosting revenue as much as possible from market segments willing to pay a high price, a corporation can make fewer but more profitable sales by pricing a new product high.

(ii) **Market-penetration pricing:** Bringing down the cost of a new product in order to expand its market and client base.

5.1.4.2.2 Product Mix Pricing Strategies

(i) **Product line pricing:** The pricing differences between different products in a product line are determined by the cost differences between the goods, client evaluations of various characteristics, and competitors' prices.

(ii) **Optional-products pricing:** The price of supplemental, ancillary goods that enhance a primary offering.

(iii) **Captive-product pricing:** pricing accessories like razor blades and camera film that must be purchased in addition to the main product.

(iv) **By-product pricing:** Pricing by-products to increase the main product's price's ability to compete.

(v) **Product bundle pricing:** combining numerous items and charging less for the package.

5.1.5 Issues in Price Management in Pharmaceutical Industry

1. **Pricing over the life cycle of the product:** Each product has a unique life cycle, and over time, its sales and profitability fluctuate. The pricing methods and issues must change as the product moves through each of these stages. For instance, the introduction (or launching) phase, which is marked by relatively low sales levels, a moderate pace of sales growth, and little or no profitability. The following two characteristics define the second growth phase: revenue growth that is accelerating quickly and significant increases in product profitability.
2. **The rate of market growth:** Price-setting practices frequently have an impact on this. In reality, certain things have a lower inherent chance of

gaining a quick market than others. Therefore, a pricing strategy that starts with low and maybe even negative margins is unacceptable for these products.

3. **The erosion of distinctiveness:** How fast originality is lost depends on how many rival products are introduced to the market and their capacity to duplicate the attributes of pioneer products.
4. **The significance of cost:** One of the crucial components is the producers' cost structure. Real cost reduction may be influenced by measuring the cost for a later period. The volume of output during that time period would have an impact on how much would be reduced.
5. **Post-skimming strategies:** The practice of charging a product's highest possible starting price is known as price skimming. As the first customers' demands are met, the business lowers the price to entice a second, potentially more price-sensitive segment.
6. **Mixed strategies:** Many businesses do choose a middle ground between the skimming and penetration price trajectories. Many large businesses adhere to this method while introducing numerous new items. For instance, **Du Pont** uses a hybrid approach to science that deals with innovation and healthy living, such as the use of nylon and cellophane.

5.1.6 An overview of DPCO (Drug Price Control Order) and NPPA (National Pharmaceutical Pricing Authority)

5.1.6.1 Drug Price Control Order (DPCO)

The objective of DPCO is to achieve adequate production and regulate equal distribution in industries. It also maintains and increase supply of bulk drugs and formulations and control the fair pricing.

DPCO is a government order that gives it the authority to set a selling price for essential and life-saving medications (in line with a predetermined formula) in order to make sure that the general public can access these treatments for an affordable price. This is done with the help of the power granted by Section 3 of the Essential Commodities Act of 1955. On May 15, 2013, the most recent Drug Price Control Order (DPCO-2013) was released.

Medications listed in Schedule I of the Drug Price Control Order (DPCO), which is periodically published by the Government of India, are referred to as "scheduled drugs" or "scheduled formulations."

Since 2013, the government has designated particular formulations as "Essential Medicines" by publishing a National List of Essential Medicines (NLEM). The NLEM-2011 list is included in the DPCO-2013 Schedule I. Therefore, NLEM is used to determine which medications should be covered under DPCO price regulation. Price fixation may apply to any formulation based on a combination of these medications that is classified under NLEM.

NLEM was not taken into consideration for price fixation or price monitoring in the DPCOs that preceded DPCO-2013.

As of 2013, any crucial drug (as defined by NLEM) is now treated as a scheduled formulation. DPCO-2013 (under). This does not, however, imply that all medicines covered by price controls are essential medicines. In accordance with DPCO-2013 Paragraph 19, during the duration of the relevant time, any drug's retail price or ceiling price may be determined by the government, regardless of whether it is a new drug or one that is not on the schedule.

Additionally, it has the authority to change the cap price or retail price of a drug that has already been established and made public, independent of the annual wholesale price index for that year (based on which companies are automatically permitted under DPCO to revise the prices annually).

Price controls are applicable irrespective of whether it is generic or branded.

Retail price calculation under DPCO: According to DPCO rules, the retail price of a product is determined in the manner described below:

Retail Price (R. P) = (M. C+C.C+P.M+P.C.) x (1+ MAPE/100) +ED.

M.C = Material cost, **C.C** = Conversion cost, **P.M** = Packaging material cost **P.C** = Packing charges, **E.D** = Excise duty and **MAPE** = Maximum allowable post manufacturing expenses

Penalties on **Offences not following the DPCO**

- Retail price shall not to exceeded or extra local taxes should not be exceeded beyond the limit mentioned under DPCO.
- If someone is responsible for finding something wrong, they will be sentenced to one year in prison.

5.1.6.1.1 Recent steps taken by the government to control Drug Prices in India

1. **Reduction of prices of medicines:** About 33 necessary pharmaceuticals had their costs reduced by the National Pharmaceutical Pricing Authority, which regulates drug pricing, in September 2016, resulting in a 30–50% drop in their retail prices. Antibiotics, pharmaceuticals for ulcerative colitis, antihistamines for colds and coughs, meds for arthritis, GERD, psoriasis, and tuberculosis are some of these.
2. **Reduction in the prices of coronary stents:** Coronary stents were included in the National List of Essential Medicines, 2015, by the Ministry of Health and Family Welfare in July 2016. (NLEM, 2015).

   The National Pharmaceutical Pricing Authority (NPPA) issued an order on February 13th, 2017, capping the price of coronary stents at a reduction of up to 40% from their current market costs.

5.1.6.2 National Pharmaceutical Pricing Authority (NPPA)

Under the Drugs (Pricing Control) Order of 1995, the National Pharmaceutical Pricing Authority (NPPA), a government agency in India, is responsible for fixing or revising the prices of bulk drugs and formulations that are controlled, as well as for enforcing drug prices and availability. The NPPA was established on August 29, 1997. The NPPA frequently releases a list of medications along with their highest ceiling pricing. The most recent DPCO, which includes a list of 384 medicines, was published in 2013.

Functions of NPPA

- In line with the authority provided to it, it executes and enforces the conditions of the Drugs (Prices Control) Order, handles all legal matters arising out of the Authority's decisions.
- The NPPA keeps an eye on drug availability, identifies any shortages, and takes corrective action. It gathers and maintains data on manufacturing, exports and imports, individual company market shares, company profitability, etc., for bulk medications and formulations.
- Conduct relevant studies related to drug/pharmaceutical price and/or sponsor them.
- It hires/appoints the Authority's officers and other staff members in accordance with the regulations and guidelines established by the government.
- Provides suggestions to the central government for changes to the drug policy.
- Assists the Central Government in parliamentary proceedings involving drug pricing.

5.2 Emerging Concepts in Marketing

Marketing is a sort of communication that involves passing along ideas and information about a product or service from the source to the recipient. A solid selling of products using promotional tools like advertising, PR, media salesmanship, etc. is also included. Both vertical and horizontal marketing are used in the marketing system.

5.2.1 Vertical & Horizontal Marketing (Table 5.1)

Vertical Marketing System: Is to attract and get in touch with businesses in the same sector. The vertical marketing system consists of the producer, wholesaler, and retailer, and the wholesaler's role is to purchase the products from the producer and deliver them to the retailer. The cooperation of these three aims to increase profit. For instance, a producer of water pumps may market its goods to installers and other professionals who use water pumps.

Comparable-sized businesses collaborate using this marketing method to take advantage of economies of scale. The companies pool resources including marketing, distribution, production, and even human resources in an effort to boost their profits. Increasing market competition has led to a recent rise in popularity for this marketing tactic.

Objectives of vertical marketing:

- By taking away the margins that are often held by third parties, it significantly lowers production costs. Ensure supply continuity and product quality to gain a competitive edge.
- Boost negotiation skills and enhance work planning.
- Improve process control; lessen the requirement for a sizable staff and a complex organisational structure. Consequently, by relying less on third parties, it is possible to concentrate on providing value to the clients.

Benefits of vertical markets:

(i) The capacity to concentrate all marketing efforts on a very particular user or consumer for optimal efficacy.

(ii) Acquired experience and knowledge in a certain niche setting, which eventually helps to boost success and market exposure.

(iii) Marketing methods are more expensive than more efficient marketing initiatives.

(iv) The chance to provide specialised goods and services that can be priced higher due to market authority and proficiency.

(v) The opportunity to establish trusting bonds with customers, suppliers, and distributors.

(vi) The advantage is less competition, which might result in speedier sales.

Example of Vertical marketing system: ***Amway*** is a multi-level marketing firm based in the United States that sells home care, personal care, and health products. Amway develops its own line of products and exclusively distributes them through registered Amway stores. Here, the firm itself is in charge of manufacturing and distribution.

Horizontal marketing system: An organization with a horizontal marketing structure allows manufacturers, distributors, and retailers to collaborate. The advantage of the horizontal marketing method is that it focuses on broad audiences rather than specific niches. Additionally, it saves time because the generation of the content requires less work because it is not customized.

Benefits of horizontal market:

(i) It has the capacity to save expenses by pooling resources with other businesses.

(ii) It makes it easier to determine and improve the quality of goods or services.

(iii) The staff is more skilled and specialised, which can boost productivity levels.

(iv) It is less likely to be affected by a decline in demand because it has a larger consumer base.

(v) Offers the chance to utilise several pricing models based on various values for various markets.

Examples of Horizontal Marketing: ***Johnson & Johnson,*** a healthcare company and Google have partnered with the goal of creating a platform for robotically assisted surgery. This will facilitate the incorporation of cutting-edge technologies and enhance healthcare services. Consequently, two or more businesses collaborate in order to take use of each other's knowledge and increase their market share.

Table 5.1 Differences between Vertical and Horizontal Marketing Systems

S.No.	Features	Vertical Marketing	Horizontal Marketing
1.	**Definition**	Aims to draw in and contact similar companies.	A marketing strategy wherein comparable businesses collaborate to achieve economies of scale.
2.	**Players**	includes producers, distributors, and retailers, all of whom collaborate with the goal of maximising profits.	It includes either a manufacturer, wholesalers or retailer.
3.	**Demographic**	Make an effort to connect with a certain population.	It appeals to a broad, non-specific demographic.
4.	**Partnership opportunities**	Don't foster a partnership culture.	Provide more opportunities for partnerships.

5.2.2 Rural Marketing

Rural marketing is the process of creating, pricing, promoting, and distributing goods and services that are specialized to rural areas in order to fulfil organizational goals as well as the intended exchange with rural clients.

Or, ***Rural Marketing*** refers to the actions made by marketers to persuade individuals living in rural areas to use their purchasing power to effectively demand the goods and services that are accessible there, improving their level of living and attaining the company's overall goal.

Rural area is a region that lies outside of towns and cities. According to the census, if less than 400 people live there per square kilometer, a location is deemed rural.

There is a two-way procedure used in rural marketing, such:

Urban to Rural: The urban market offers the rural market fast-moving consumer goods (FMCG), agricultural fertilizers, vehicles, etc.

Rural to Urban: Fruits, vegetables, flowers, milk, and other agricultural products are offered from the rural market to the urban market.

To understand the difference between ***Rural and Urban*** market, the following **Table 5.2**. is depicted below.

Table 5.2 Difference between Rural and Urban Marketing

Basis of Difference	Urban	Rural
Product	Product Large packaging, more emphasis on quality	Small packaging, More emphasis on core benefit
Place	Direct and Indirect Distribution Channel	Indirect Distribution channel
Price	High Price Product	Low Price Product
Promotion	More emphasis on media	Emphasis on organizing events, Opinion Leader

Rural markets have become the new goals for business organizations as a result of the urban market being oversaturated with competition and being close to saturation. the elements in Figure 5.1 that have enabled rural markets. **Table 5.3** shows the advantages and difficulties of the rural market.

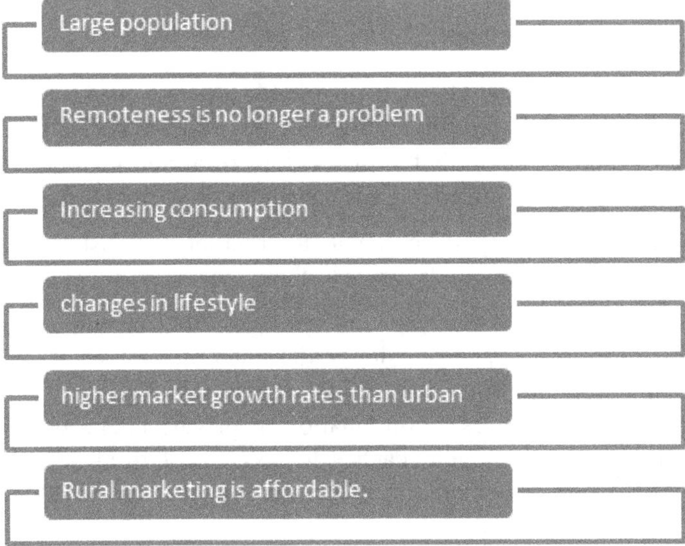

Figure 5.1 Factors make rural markets viable

Table 5.3 Benefits and difficulties of rural marketing

S.No	Benefits of rural marketing	Difficulties in rural marketing
1.	Size of rural market	Lack of proper communication
2.	Employment	Distribution problem
3.	Better living	Low literacy level and lack of adequate transport facilities
4.	Contribution to national income	Lack of adequate qualified doctors
5.	Increase in farm income	Lack of proper retail outlets

Model in rural marketing: Rural marketing is a "4A" model as given in **Figure 5.2.**

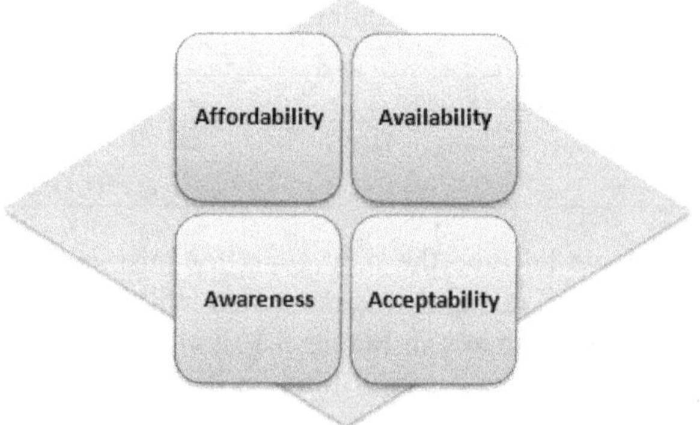

Figure 5.2 Rural marketing model

Affordability, due to their limited disposable money, it can be difficult for rural residents to assure the affordability of the good or service. For instance, Coca-Cola offered a 200 ml bottle at Rs. 5 in rural areas as an illustration of how products need to be affordable for consumers who are on daily wages.

Availability is the largest challenge for retailers or customers who come from remote villages. The goods arrive to the clients there much later.

Awareness aims to the way of reaching the customer through the commercials from media as television, radio and outdoor. Since, lack of media source in rural areas are limited affects the marketing of a product. Example, A Coca-Cola brand when uses a combination of private media presentations on T.V., cinema and radio which reaches to 53.6 % of rural households while when advertised through Doordarshan (which is a govt. media) alone it reaches to 41% of rural households.

Acceptability in rural market is based on the customer who should feel that the product is designed as per their needs and the product must be familiar to them for example, LG Electronics.

5.2.3 Consumerism

Consumerism is a consumer backlash against unethical company practices and sectors. It aims to eradicate unethical marketing strategies such false advertising, hoarding, black marketing, short masses and measures, adulteration, artificial (made) pricing, misleading packaging, false and misleading ads, unsafe products, and misbranding.

Consumerism is a concerted effort by people, organizations, and governments to safeguard consumers from laws and business practices that violate their right to fair dealing. It is a phenomenon in society that gives buyers and consumers more power. It maintains an eye on the businesses and guarantees that customers receive high-quality, safe items at the right price.

Importance of consumer protection: The goal of consumer protection is broad; it involves informing consumers about their rights and obligations and assisting them in having their complaints resolved. It is intended to protect consumer interests, but it also calls for the consumers to band together and create consumer associations to advance and defend their rights. However, the value of consumer protection to business is also particularly important.

One way to accomplish the goals of consumer protection is by ensuring that consumers are aware of their rights and obligations.

Types of consumerism in marketing: When a company markets a product or service, it is essential to comprehend how consumers decide what to buy. In other words, the company's marketing plan is influenced by the purchasing habits of the public. Chewing gum is a common example of a product that is advertised on a display near the checkout counter so that customers can pick up a pack as they leave. This type of product is referred to as a "impulse purchase." To compare products, though, and to learn more about the features and advantages of a product before making a significant purchase. The following four forms of marketing consumerism could be regarded as:

1. **Routine purchases:** People occasionally make purchases that don't need much thought. The consumer's "Programmed behavior" leads them to make these purchases. This is due to the fact that the buyer makes very little effort to find the goods and decide which product to purchase when making this type of purchase. Additionally, regular purchases are generally low-cost things that don't demand a lot of the customer. Simple food products like milk, eggs, and snacks are examples of purchases that are made on a regular basis.
2. **Purchases with little amount of decision making:** Consumers frequently purchase some things, whereas others are only bought on occasion. Even if the consumer is familiar with the product category, they will still investigate a new brand to learn more. For instance, the client might have bought shirts frequently in the past, but a new product made of a different material might

only require the buyer to gather a modest amount of information. Additionally, the customer will take some time to research a new brand.

3. **Purchases with high amount of decision making:** Consumers spend a significant amount of time researching and choosing what to buy. These goods are typically unknown, pricey, and occasionally purchased as well. When purchasing these products, the buyer takes a significant financial, psychological, or performance risk. These large purchases include, for instance, automobiles, houses, and computers.
4. **Impulse buying:** The same customer who spends hours deciding which computer to buy frequently makes other purchases without a second thought. Decisions are made on the spot and there is no mental planning involved in this impulsive buy. Typically, these products include candy bars, magazines, gum, and similar stuff.

5.2.4 Industrial Marketing

Industrial marketing is the act of one company trying to sell industrial goods or services to another. To be clear, everything that facilitates the creation of a finished good from a raw material is regarded as an industrial good or service. Industrial marketing refers to the advertising of products and services by one business to another (also known as business-to-business marketing). Companies that use industrial products to transform one or even more raw materials into a finished good are called as such. The term "industrial marketing" has virtually been replaced by "B2B marketing" (i.e. business to business marketing). Government initiatives, manufacturing firms, educational institutions, private sector organizations, distributors, dealers, and hospitals are a few of the targets in this trade. The purpose of business organizations' purchases of goods and services is to meet a variety of needs, including cutting expenses, producing goods and services, making money, etc. The B2B selling process' fundamental components include:

- Due to the one-on-one nature of marketing, it is very easy for a seller to identify a potential customer and build a personal rapport.
- Buying processes include highly qualified and skilled individuals. A purchasing plan frequently needs to receive the approval of two or three decision makers.
- The purchasing or selling procedure is frequently complicated and consists of several steps (for instance, requests for proposals and tenders, the selection and awarding of the winning bid, contract discussions, and the signing of the final agreement).
- Prospecting, qualifying, courting, presenting, drafting tenders, building strategies, and contract negotiations are all time-consuming steps in the selling process.

- Business buyers are much less numerous than consumer purchasers, but they make larger or more frequent purchases.
- The buyer-seller relationship must be maintained at all times in a professional manner. Since it indicates tremendous profit, relationships that have already been established are difficult to change. Any deal needs regular follow-ups and ongoing sales calls to be completed.
- To sell the goods to industrial buyers, sharp sales ability is needed. Since they are making large purchases, they won't be readily persuaded, therefore in-depth product knowledge is necessary.
- In order to complete the product, many authorities and officials will interfere and sway consumers' choices.

If a product is pushed using freebies, demonstrations, etc., it will require less promotional work for any sector because broad audiences won't be exposed to it. Once a relationship is established, it results in a steady flow of clients who buy your items over an extended period of time. Therefore, if current customers are appropriately satisfied, the future is assured.

5.2.5 Global Marketing

Global marketing is the execution of commercial actions intended to organize, advertise, and guide a company's flow of products to consumers in multiple countries. To conceptualize and then communicate a finished good or service globally in an effort to engage the global marketing community.

Global marketing is affected by driving forces as played by:

1. ***Technology*** which play a significant part in global marketing, for instance, the internet, mobile phones, e-mail, video conferencing, fax, satellite TV, and other technologies that have increased communication capabilities and lowered the barriers of distance, time, and cost between nations.
2. ***Heavy cost of new product*** introduction is what motivates people to enter the world market. In the pharmaceutical industry, it typically takes 7 to 10 years to launch a new medication based on research, and to recoup the enormous costs, businesses often aim to expand internationally.
3. ***Use of marketing strategies*** by businesses having advantages in one nation compared to another, with any necessary revisions.
4. ***Guaranteed performance,*** if a corporation can deliver a product in an international market, it'll be willing to sell many units and earn a big profit.
5. ***Global scale manufacturing facility***, the business can promote its goods in multiple nations and charge a reasonable price for them. Although there are few worldwide marketplaces, they have been developed by marketers. Numerous nations now have a market for soft drinks thanks to businesses like Pepsi and Coca-Cola.

When done well, global marketing has many advantages.

- To start, it can increase a product or service's effectiveness. This is due to the fact that when a product grows, a person learns more about the efficient production of new goods and services.
- Having a sizable competitive edge is the second. It is easy enough for enterprises to compete in the local market. But relatively few companies are able to compete on a worldwide level. Therefore, if a company can compete successfully, competitors cannot exist, and the company becomes a dominant force globally.
- Third, a rise in consumer knowledge of the company, its name, and its goods or services. Consumers can follow global development online using the internet.
- Global marketing can also enhance savings by lowering costs. One can get scale economies and range by standardizing the procedures and focusing on various markets, in addition to the cost savings from using the internet.

Global marketing entry strategies:

A. **Exporting:** Operating in overseas markets is a comparatively low risk strategy. It is possible to describe exporting as the marketing of domestically produced goods for sale abroad. Exporting has a number of benefits and drawbacks, which are listed in **Table 5.4.**

Table 5.4 Advantages and disadvantages of exporting

S. No	The Benefits of Exporting	Implications of Exporting
1.	Manufacturing unit is home based so it is less risky than other strategies.	Companies' sales are based on the dealing of foreign agents.
2.	Offer company an opportunity to learn the foreign markets before directly investing in foreign countries.	Lack of control on sales because of dependency on foreign agents
3.	Enables full control on production unit	Company cannot expect for very high profits.

B. **Licensing:** "Licensing is the means of accessing the global market," says Philip Kotler, "whereby a corporation establishes a deal with a foreign company to offer licence for manufacturing, processing, and selling the product in return for royalty." For instance, the Chinese corporation "Tasly" has granted Torrent Pharmaceuticals a licence to sell its cardiovascular drugs. Exporting has a number of benefits and drawbacks, which are shown in **Table 5.5.**

Table 5.5 Advantages and disadvantages of licensing

S.No	Advantages of Licensing	Disadvantages of Licensing
1.	Little expenses are required because only cost is involved in signing the agreement.	Control is lost on profit sales so big profit cannot be expected.
2.	The efforts of two companies are merged for better output.	Company sales depend on performance of other company that receives license.
3.	Less involvement and efforts are required by company providing license.	Sometimes license receiving company can use low quality ingredients that can make bad image of license providing company.
4.	Company gets provision to take royalties in stock which can be invested elsewhere.	The partners company (license receiving company) will know the technology of main company and can become a competitor of main company in future.

C. Contract Manufacturing: In this strategy a company makes tie up with some other company to manufacture the product but governs it promotion, pricing and selling on its own. It has its own advantages and disadvantages as listed in **Table 5.6.**

Table 5.6 Advantages and disadvantages of contract manufacturing

S.No	Advantages of Contract Manufacturing	Disadvantages of Contract Manufacturing
1.	Company can start with less economic resources.	Profits are shared between two companies.
2.	Lesser risk than direct investment (Direct entry strategy).	It involves indirectly training a future competitor.
3.	The strains of manufacturing of product is avoided.	
4.	The combined efforts of two companies are merged together.	

D. Joint ownership: It is a form of joint venture having advantages and disadvantages as mentioned in **Table 5.7**. In this two companies equally share the ownership and control over the property rights and operation and combined to invest in market of foreign country.

Table 5.7 Advantages and disadvantages of Joint ownership

S.No	Advantages of Joint ownership	Disadvantages of Joint ownership
1.	Joint financial strength.	Profits are divided between two companies.
2	Risk is also shared between both companies equally.	Partners may have different views on technology policy and expected benefits.

E. **Direct investment strategy/ Direct Entry Strategy:** The strategy where company directly manufactures, makes policie's, sells and governs all the foreign related business on its own. So company bears all investment by itself and it also have few advantages and disadvantages as given in **Table 5.8.**

Table 5.8 Advantages and disadvantages of Direct investment strategy

S.No	Advantages of Direct investment strategy/ Direct Entry Strategy	Disadvantages of Direct investment strategy/ Direct Entry Strategy
1.	Company can expect high profits.	Large investment is required at once.
2.	It has full control on foreign sales.	Risk is owned by company fully.
3.	It is fully responsible for all decision.	
4.	All profits are enjoyed solely.	

Global marketing policies have been designed for:

A. **Global product policy:** Generally, there are some major policies adopted by companies for global marketing. They are as follows:

(a) **Product extension**: Without altering the product or its promotion efforts, the company markets its products abroad.

(b) **Communication adaptation**: Instead of changing the product, the company uses a different communication approach in the international market. For instance, Colgate devised a separate advertising strategy for India, the United States, and Indonesia since white teeth are favored in the former while yellow teeth are preferred in the latter.

(c) **Product adaptation**: The business adapts its offerings to local preferences while maintaining the same communication practices. Due to the prevalence of vegetarianism in India, KFC began selling vegetarian burgers there even though its non-vegetarian brand is more well-known.

(d) **Dual policy:** This policy changes both the product and the communication in accordance with the demands of the local market. For illustration, Nokia expanded their cell phone volume choice in India (Because most of the places in India are overcrowded). Additionally, a lot of people in India do not speak English, therefore Nokia modified their communication approach in line with regional differences.

(e) **Product invention**: The business creates a totally new product to satisfy local clients' needs. Examples: Nokia created the "Nokia-1100 set" exclusively for the Indian market (Because most of Indians believe in durability and low prices).

B. **Global promotional policy**: Promotion in global market is quite challenging because of existence of various language, dialects, culture, traditions and perceptions. Similarly, there are strict regulations on the advertisement and sales promotion schemes (gifts, discounts, free vouchers etc.) in some countries. So, there must be adequate standardization of promotions policies according to requirement of various countries. Example: If a company markets in Japan, it should not use the white color for packing or background because they consider white only for mourning.

C. **Global branding policy**: The brand name of a product must be developed carefully after keeping in mind its global acceptance. Since, brand name of a product that can't be changed according to local variations, so they must be named very cautiously. Brand name should have specific attributes as small and easy to pronounce, proper relevance with the product or its features. It must appeal to consumers and provide the stimulation they need to buy the products..

D. **Global Pricing Policy:** The pricing decision is very critical in global marketing because it affects the sales and profit of product. The price varies from one region to another due to several factors as:

- Cost of distribution in various areas with transportation cost
- Availability of competitors varies in different areas as per the demand of product which can vary in from place to place.
- Product brand image may be good, average or low in different areas. There must be the substitute products availability.
- As Government regulations differ from country to country the extra costs like insurance payments, export cost, packaging cost, documentation cost, taxes can differ.
- Economic condition of people living in the country.

Hence, the company follows different pricing policy in different regions for optimizing its sales and profits.

Constraints to Global marketing: Although Global marketing seems to be profitable, there are several barriers in the way of global marketing:

1. **Language barrier:** The language differs from region to region, so before starting foreign business advertising and labeling of products has to be translated according to local language.
2. **Cultural barrier:** The cultural difference may make an advertisement performing well in one nation, totally unsuitable in another. Example: A new camera which was launched in US and Japan show different cultural variability due to the advertisement. It can be due to the advertisement which was doing well in Japan but in US objections were raised against the advertisement.

3. **Local attitude toward advertising:** People in some countries are more receptive to advertising than others. Example: Advertising is considered as a fact of product launching in US but Europeans think that advertising is only for making money and can be false claiming also.
4. **Media Infrastructure:** In some area TV media is less developed or people tend to favor reading of newspaper more than TV. Hence advertising should be designed according to the media habits of the country.
5. **Advertising regulations of Government:** Different countries have different regulations related to advertising. Some developed countries are broader minded in advertising special products (condoms) while such sort of advertisements is banned in many Asian countries. Similarly, comparative advertising is banned in some countries but it is allowed in US.

Global Pharmaceuticals Market: At a compound annual growth rate (CAGR) of 1.8%, the worldwide medicines industry is anticipated to increase from $1228.45 billion in 2020 to $1250.24 billion in 2021. In 2025 and 2030, respectively, the market is anticipated to reach $2,050.9 billion and $3,206.3 billion. The COVID-19 pandemic, changes in lifestyle, new techniques for drug discovery, a sizable pool of undiagnosed population, and a rise in the use of pharmaceutical drugs will all propel the market's expansion. Reduced free trade and high production costs for biologics are two factors that could limit the expansion of the pharmaceuticals sector.

With 60% of the world's vaccinations originating from India, Indian pharmaceutical businesses have built a name for themselves internationally. Indian pharmaceutical companies control more than 20% of the US prescription market, and India has the most FDA-approved production facilities outside of the US.

Businesses (groups, sole proprietorships, and partnerships) that create and market drugs used to treat ailments as well as related services make up the pharmaceuticals market. Any type of medication used to treat ailments or for medical purposes can be referred to as a pharmaceutical. This industry includes companies that produce pharmaceuticals and biologics. The two pharmaceuticals market segments are pharmaceutical drugs and biologics.

Answer the Following Questions

Q1. Enumerate the importance of pricing in marketing.

Q2. Include pricing goals in your list of considerations.

Q3. What are the major Determinants of price of a product?

Q4. Discuss the pricing methods used in marketing.

Q5. Discuss the Issues in price management in pharmaceutical industry.

Q6. Describe the most recent actions the government has taken to control drug prices in India.

Q7. Explain various pricing methods for inventories.

Q8. Enlist the functions of National Pharmaceutical Pricing Authority.

Q9. Create the process for fixing or changing the pricing of bulk medications.

Q10. Compare and contrast the vertical and horizontal marketing systems in your writing.

Q11. Define rural marketing and its objectives.

Q12. Enlist the problem and difficulties faced in rural marketing.

Q13. Write a note on 4 A Model in rural marketing.

Q14. Compare and contrast rural and urban marketing.

Q15. Discuss the importance of consumer Protection.

Q16. Illustrate the various constraints of Global marketing.

Q17. Comment on various policies which are adopted by companies to push the product in global market.

Q18. Write a note on various driving forces which are significant for Global marketing.

Q19. Define industrial marketing. Enlist the feature of B2B selling process.

Q20. Differentiate between rural and urban marketing.

MCQs

1. Which of the following significantly affects price choices?
   A. Customer demand
   B. Actions of competitors
   C. Costs
   D. Political, legal, and image-related issues
2. Which, in marketing speak, means "what we receive for what we pay"
   A. Revenue B. Cost
   C. Value D. Product
3. Who has the authority to determine the retail price for scheduled formulations?
   A. State Government B. Central Government
   C. Lok Sabha D. Rajya Sabha

4. Any violation of the Drug (Prices Control) Order's provisions is punishable by the provisions of the
   A. Drugs and Cosmetics Act, 1940
   B. Narcotic Drugs and Psychotropic Substances Act, 1985
   C. Essential Commodities Act, 1955
   D. Industries (Development and Regulation) Act, 1952

5. According to DPCO, each factory is required to submit yearly data on turnover and the distribution of revenues and expenses in
   A. Form11 B. Form 1V
   C. Form V D. Form V1

6. The manufacturer is required by D P C O to file an application to change the selling price for a scheduled formulation in
   A. Form Vl B. Form V
   C. Form 111 D. Form 11

7. NPPA (National Pharmaceutical Pricing Authority) comes under
   A. Ministry of health
   B. Ministry of commerce
   C. Ministry of Chemical and Fertilizers
   D. Ministry of Finance

8. Extension of marketing efforts globally is referred to as
   A. International Business B. Universal Marketing
   C. International Marketing D. Borderless Marketing

9. Which factors affect international marketing decisions.
   A. Political B. Economical
   C. Social D. All of the above

10. The only distinction between domestic and international marketing definitions is
    A. The marketing activities take place in more than one country.
    B. The marketing activities take place in one country only.
    C. The marketing activities take place in host country only.
    D. The marketing activities must take place in all countries.

11. The best way to categories a detailed stated version of shortlisted new ideas in relevant consumer terms is
    A. Concept B. Phase
    C. Screening D. Raw-material screening

12. The propensity for challenging comprehension associated with the utilization of market offerings is known as

    A. Relative advantage B. Complexity
    C. Communicability D. Compatibility

13. Rural marketing that is organized through can be more successful:

    A. Door- to -door campaigns B. Melas
    C. Village fairs D. All of the above

14. Marketers determine the pricing.

    A. Select the pricing objective
    B. Estimate demand
    C. Analysis competitors' cost, offers and prices
    D. All of the above

15. The pricing objectives are

    A. Maximum current profit, market share and market skimming
    B. Survival
    C. Product quality leadership
    D. All of the above

16. Walkers brand is used in

    A. UK B. Australia
    C. USA D. China

17. The strategy of a company which makes tie up with other companies to manufacture the product is by

    A. Direct entry B. Contract manufacturing
    C. Licensing D. Exporting

18. Torrent Pharmaceuticals is authorized to offer Chinese company's cardiovascular medications.

    A. Tasly B. Sislay
    C. Werley D. Titley

19. Among the following which is odd with respect to industrial marketing management.

    A. Product exchange B. Information exchange
    C. Financial exchange D. Buyer exchange

20. The phrase "——— marketing" has largely taken the place of the phrase "industrial marketing."

    A. B2B B. B2C
    C. C2C D. C2B

21. AI can be expressed as
    - A. Artificial intelligence
    - B. Artificial income
    - C. Aircraft intelligence
    - D. Adaptation intelligence
22. It was approved the Consumer Protection Act in
    - A. 1986
    - B. 1982
    - C. 1984
    - D. 1988
23. Which company works with the Self-Employed Women's Association (SEWA)
    - A. Eli Lilly
    - B. Mankind
    - C. Sun Pharma
    - D. Novartis
24. Nicholas Piramal (NPIL) has tied up with Sorento Healthcare Communications for an----- Outreach Programme.
    - A. Epilepsy
    - B. Cancer
    - C. Depression
    - D. Diabetes
25. LG Electronics. In 1998, it developed a customized TV for the rural market and named it Sampoorna
    - A. Sampoorna
    - B. Devpoorna
    - C. Servpoorna
    - D. Anshpoorna
26. Marketing system that aims to attract and reach businesses operating in the same industry.
    - A. Vertical
    - B. Horizontal
    - C. Parallel
    - D. Mixed
27. In DPCO "O" stands for
    - A. Order
    - B. Objection
    - C. Official
    - D. Object
28. In July 2016, the Ministry of Health and Family Welfare entailed coronary stents in the
    - A. NLEM, 2015
    - B. NLEM, 2010
    - C. NLEM, 2005
    - D. NLEM, 2008
29. Revenue-oriented pricing is also known as
    - A. Operations-oriented pricing
    - B. Customer-oriented pricing
    - C. Value-based pricing
    - D. Profit-oriented pricing
30. Value-based pricing
    - A. Operations-oriented pricing
    - B. Customer-oriented pricing
    - C. Image-based pricing
    - D. Profit-oriented pricing

Answer key to MCQs

1. A, 2. C, 3. B, 4. C, 5. D, 6. C, 7. C, 8. C, 9. D, 10. A,
11. A, 12. B, 13. D, 14. D, 15. D, 16. A, 17. B, 18. A, 19. D 20. A,
21. A, 22. A, 23. A, 24. A, 25. A, 26. A, 27. A, 28. A, 29. A, 30.C

www.ingramcontent.com/pod-product-compliance
Lightning Source LLC
LaVergne TN
LVHW010038160826
845671LV00003B/178
9789395039727